# Convergence

# Convergence

## Artificial Intelligence and Quantum Computing

Edited by

**Greg Viggiano, PhD**

WILEY

Published by John Wiley & Sons, Inc., Hoboken, New Jersey.
Published simultaneously in Canada and the United Kingdom.

ISBN: 978-1-394-17410-2
ISBN: 978-1-394-17412-6 (ebk.)
ISBN: 978-1-394-17411-9 (ebk.)



For general information on our other products and services or for technical support, please contact our Customer Care Department within the United States at (800) 762-2974, outside the United States at (317) 572-3993 or fax (317) 572-4002.

If you believe you've found a mistake in this book, please bring it to our attention by emailing our reader support team at wileysupport@wiley.com with the subject line "Possible Book Errata Submission."

Wiley also publishes its books in a variety of electronic formats. Some content that appears in print may not be available in electronic formats. For more information about Wiley products, visit our web site at www.wiley.com.

**Library of Congress Control Number:** 2022946537

Cover image: © Michał Klimczak
Back Cover logo provided courtesy of The Museum of Science Fiction, Washington, DC
Cover design: Wiley

SKY10037583_102822

*"Fantastic machine, the M-5. No off switch."*

*—Dr. Leonard McCoy*

*"The Ultimate Computer,"* Star Trek, *air date March 8, 1968*

# Contents

# Preface

When science fiction suddenly becomes reality, the world watches with astonished fascination, delight, and sometimes dismay. We live in an accelerated time, and the rate of acceleration is increasing. As we rapidly move forward, it is difficult to see over the horizon. Yet, it is wise to make preparations for what lies ahead. A paradox? Perhaps. The essential question is, how can one adequately prepare for the unknown?

This acceleration point may have started back in the mid-19th century. "What hath God wrought?" (a phrase from the Book of Numbers 23:23) was the first Morse code message transmitted in the United States on May 24, 1844, and officially opened the Baltimore–Washington telegraph line. The phrase was suggested to Samuel Morse by Annie Ellworth, the daughter of the commissioner of patents and appropriately called attention to an obvious, world-changing event. A harbinger of definite magnitude.

Another technological watershed is now coming into place: artificial intelligence converging with quantum computing. The convergence of these two technologies may have the same civilization-altering effects as the telegraph, but the changes resulting from their combined functionality are likely to be much more profound, perhaps as fundamental and far-reaching as the discovery of fire.

At the present time, the technology maturation path for artificial intelligence and machine learning is clearer than that for quantum computing. But, as classical computing uses more sophisticated machine learning tools to advance better and better quantum computing designs, it is not unreasonable to expect that progress will continue to accelerate, eventually even exponentially. So, what happens when continuously accelerating development of this technology is able to proceed without any limitations? One possibility may come in the form of a super-watershed where the power of the combined technologies is able to create much higher performance tools—tools that become so sophisticated that they begin to improve themselves and find solutions before we even understand the problems.

Perhaps unsurprisingly, opinions differ over the current state of the technology, depending on precisely how quantum computing is defined. For instance, some feel that quantum computing is not practically functional until certain thresholds have been achieved, i.e., a minimum number of qubits, room temperature operation, etc. For this collection, the individual authors have taken sometimes differing positions on how they define quantum computing and its current state of maturity, so there are necessarily differing assumptions in certain contributions. The intention in providing such a range of opinions is to try to bring a truly wide-angle lens to bear on the analysis of the impending revolution.

In some sense, the revolution is already underway. Look at all of the related technologies currently in development: guidance systems for autonomous vehicles and aerial vehicles, military applications, financial portfolio optimization, cryptography, network communications, medical research . . . the list gets longer each year.

Much in the same way that electricity became ubiquitous during the 19$^{th}$ century, civilization again seems to be headed down the same road with quantum-enabled AI systems. All in all, these changes may not look like a revolution—but in the beginning, real revolutions can sometimes be difficult to spot. The importance of this anthology is to develop a critical understanding of these changes and be able to see the coming revolution more clearly. With a clearer perspective, we can ideally make the right preparations. Like a tidal wave coming in slow motion, its arrival is certain, but its size remains to be seen, and the high ground is relative to our preparedness.

To be clear, this collection of essays is not meant to provide an in-depth education about the theoretical foundations of quantum computing or artificial intelligence. The central thesis of this anthology is to raise awareness of this quiet revolution. However, to provide the reader with additional technical background should it be required, two primers on the foundational concepts of quantum computing and artificial intelligence are included in the appendices. This information is meant to simplify and explain the current state of the technologies discussed in this collection. In addition, a glossary of common definitions is provided at the end of this volume for better understanding of the more technical terms, and an index is included for easy reference to specific information.

The volume in front of you is the first in a planned series, and this installment specifically explores the potential impacts on people from AI converging with quantum computing. As with the introduction of any higher performance tool, humanity adopts the innovation and soon becomes more efficient. Left unchecked, the adoption and increased efficiency usually carry certain consequences in the form of social, economic, and political adjustments, and it is these adjustments that the current volume will investigate.

Next in the series, Volume 2 will be concerned with a full range of potential applications and use cases for the technology across various industry sectors. By understanding how the combined technologies might actually be deployed, the reader can gain a sense of where and how the way we live will be transformed (or even cease to exist). Volume 2 is meant to be an early warning signal for those likely to be affected in the first wave.

Volume 3 will build on the awareness gained from learning about the various applications and use cases and will discuss potential vulnerabilities and dependencies in need of protection and fail-safes. Without having adequate controls for disaster recovery and manual overrides, the potential to avert runaway trains will be greatly diminished or eliminated.

We are truly in the pre-acoustic coupler days (to use an ancient telecommunications reference), and the early stages of quantum-enhanced AI systems are still a few years away. But like PCs in the early 1980s, hybrid architectures will soon emerge to improve performance—similar to the 386/387 math coprocessors used to speed up complicated spreadsheet calculations. Eventually, multicore processors became fast enough to do everything on their own—including full-motion video that we take for granted today. The same development path will likely happen for quantum platforms and AI systems: classical architectures will be used for handling data-heavy tasks, and quantum (co)processors will be used for dealing with very complex calculations.

It is inherently difficult to predict how a technology will develop and mature at such an early stage of its lifecycle. The permutations will likely bear little resemblance to the tools we use today. Nonetheless, it is important to attempt an understanding of how these changes may evolve so preparations can be made and unpleasant surprises can be minimized. We may never be able to fully prepare for what may come from artificial intelligence converging with quantum computing, but we do have a little time to think about the possibilities. Thought experiments, symposiums, and game theory exercises may help extend our ability to anticipate the unexpected and see a little further over the horizon.

Given the rapid development of both technologies and given their eventual convergence, this anthology's central question is, how will this combined technology affect civilization? To help shed light here, 26 international authors were asked to speculate on the impacts of artificial intelligence converging with quantum computing. These authors were selected to achieve a multidimensional balance across geography, gender, ethnicity, professional area, and individual outlook. Their backgrounds and viewpoints raise awareness of the socio-economic, and political-regulatory impacts and describe unexpected societal changes and what may be in store for humanity.

The essays in this anthology are organized into three sections and examine the potential global impacts on political/policy/regulatory environments, economic activity, and social fabric. These impacts are complex in nature, and while there may be some degree of overlap between sections and across the individual essays, the positions presented by the authors are intended to provoke thought and consider possible consequences.

Quickly understanding the competitive advantages of using a new tool has always ensured dominance in commercial and geopolitical environments. Frequently, these advantages have strategic military capabilities for enhancing national control and global supremacy. The nations that control these tools will be able to secure their position and dominate those without the same capabilities. Quantum computing is the newest tool in this arsenal. When combined with artificial intelligence, a quantum computer can potentially solve very complex national problems, such as resource allocation, or global problems, such as climate change. Alternatively, the tool can be weaponized just as easily and applied to decrypting national security information and gaining access to military control systems.

Global commercial systems are almost always affected by the introduction of new tools and technologies, and this dimension is considered in the second section of the book. New technologies provide competitive advantages and disrupt the way industries normally operate, and one obvious area where this advantage and disruption will first emerge might seem to be human capital and labor. However, we are already witnessing how classical AI is having a major impact in this area, with further significant disruption predicted in the near term. There is valid concern that classical AI has the potential to make a very large number of workers redundant as these workers are replaced by intelligent automated systems—potentially leaving workers to continually retrain from one type of "sunset job" to another—but the brunt of these impacts are almost certain to be felt long before AI finally converges with quantum computing. For this reason, specific examples of how quantum artificial intelligence might eventually affect labor will be considered in the second volume of the series: applications and use cases.

Other key areas of commerce that will be affected are the global financial system and market trading. Even though we already see classical AI deployed widely in these areas, as we do with labor, there remain crucial aspects to the global financial ecosystem upon which the convergence of AI with quantum computing will have a truly seismic effect. When information security is considered in this context, the situation may initiate a new sort of arms race—which directly leads into the third section of this anthology, global policy and the regulatory environment.

When new tools are introduced into an existing social system, how that social system changes and adapts has both positive and negative outcomes. This anthology presents both optimistic and less optimistic perspectives regarding this type of technology introduction. As seen with the debut of the smartphone, the near-term social impacts have been obvious and well studied, but the longer-term impacts, even 25 years after first use, remain to be seen. The essays in this anthology aim to explore the question of how quantum computing and AI, like the smartphone, may evolve and affect humanity over the coming decades, offering various perspectives on the possible outcomes.

In the longer term, as with other essential technologies, I think that the aggregate effects will be irreversible—imagine trying to live today without electricity, mobile phones, or the Internet. In spite of climate change and the current pandemic, if we are to survive as a species, optimism and careful planning will serve us well. Science fiction narratives can also provide useful guidance for speculating about future technology trends and possible trajectories—and what should be avoided. Unfortunately, this is not a thought experiment: we have already lit the fuse, and the accelerant is qubits. The future will be arriving before we know it.

As with games of chance, excitement lies in not knowing the outcome. Let us hope that as we learn more about the future of these two technologies, random chance will operate in our favor . . . and perhaps hacking the lottery with a quantum processor will become commonplace.

GRV

# Foreword

## Essential (and Mostly Neglected) Questions and Answers About Artificial Intelligence

David Brin *Author and Scientist*

This essay builds upon an earlier version first published in *Axiom* Volume 2 Issue 1.

For millennia, many cultures told stories about *built-beings*—entities created not by gods but by humans. These creatures were more articulate than animals, perhaps equaling or excelling us, though not born-of-women. Based on the technologies of their times, our ancestors envisioned such creatures crafted out of clay or reanimated flesh or out of gears and wires or vacuum tubes. Today's legends speak of chilled boxes containing as many submicron circuit elements as there are neurons in a human brain . . . or as many *synapses* . . . or many thousand times more than even that, equalling our quadrillion or more *intra*cellular nodes . . . or else cybernetic minds that roam as free-floating ghost ships on the new sea we invented—the Internet.

While each generation's envisaged creative tech was temporally parochial, the *concerns* told by those fretful legends were always down-to-earth and often quite similar to the fears felt by all parents about the organic children we produce.

*Will these new entities behave decently?*

*Will they be responsible and caring and ethical?*

*Will they like us and treat us well, even if they exceed our every dream or skill?*

*Will they be happy and care about the happiness of others?*

Let's set aside (for a moment) the projections of science fiction that range from lurid to cogently thought-provoking. It is on the nearest horizon that we grapple with matters of *policy*. "What mistakes are we making right now? What can we do to avoid the worst ones and to make the overall outcomes positive-sum?"

Those fretfully debating artificial intelligence (AI) might best start by appraising the half-dozen general pathways under exploration in laboratories around the world. While these general approaches overlap, they offer distinct implications for what characteristics emerging, synthetic minds might display, including (for example) whether it will be easy or hard to instill human-style ethical values. We'll list those general pathways in the following paragraphs.

Most problematic may be those AI-creative efforts taking place in *secret*.

Will efforts to develop sympathetic robotics tweak compassion from humans long before automatons are truly self-aware? (Before this book went to press, exactly this scenario emerged: a Google researcher publicly declared that one of the language programs he dealt with had become fully self-aware . . . the first of what I call the *robotic empathy crisis*.)

It can be argued that most foreseeable problems might be dealt with in the same way that *human* versions of oppression and error are best addressed—via reciprocal accountability. For this to happen, there should be *diversity* of types, designs, and minds, interacting under fair competition in a generally open environment.

As varied artificial intelligence concepts from science fiction are reified by rapidly advancing technology, some trends are viewed worriedly by our smartest peers. Portions of the intelligentsia—typified by Ray Kurzweil[1]—foresee AI, or artificial general intelligence (AGI), as likely to bring good news and perhaps even transcendence for members of the Olde Race of bio-organic humanity 1.0.

Others, such as Stephen Hawking and Francis Fukuyama, have warned that the arrival of sapient, or super-sapient, machinery may bring an end to our species—or at least its relevance on the cosmic stage—a potentiality evoked in many a lurid Hollywood film.

Swedish philosopher Nicholas Bostrom, in *Superintelligence*,[2] suggests that even advanced AIs who obey their initial, human-defined goals will likely generate "instrumental subgoals" such as self-preservation, cognitive enhancement, and resource acquisition. In one nightmare scenario, Bostrom posits an AI that—ordered to "make paperclips"—proceeds to overcome all obstacles and transform the solar system into paper clips. A variant on this theme makes up the grand arc in the famed "three laws" robotic series by science fiction author Isaac Asimov.[3]

Taking middle ground, Elon Musk has joined with Y Combinator founder Sam Altman to establish *OpenAI*,[4] an endeavor that aims to keep artificial intelligence research—and its products—open-source and accountable by maximizing *transparency* and *accountability*.

As one who has promoted those two key words for a quarter of a century, I wholly approve.[5] Though what's needed above all is a sense of wide-ranging perspective. For example, the panoply of dangers and opportunities may depend on which of the aforementioned half-dozen *paths to AI* wind up bearing fruit first. After briefly surveying these potential paths, I'll propose that we ponder what kinds of actions we might take now, leaving us the widest possible range of good options.

## Major Category 1: AI Based Upon Logic, Algorithm Development, and Knowledge Manipulation Systems

These efforts include statistical, theoretic, or universal systems that extrapolate from concepts of a universal calculating engine developed by Alan Turing and John von Neumann. Some of these endeavors start with mathematical theories that posit AGI on infinitely powerful machines and then scale down. Symbolic representation-based approaches might be called traditional *good old-fashioned AI* (GOFAI) or overcoming problems by applying data and logic.

This general realm encompasses a very wide range, from the practical, engineering approach of IBM's Watson through the spooky wonders of quantum computing all the way to Marcus Hutter's universal artificial intelligence based on algorithmic probability,[6] which would appear to have relevance only on truly cosmic scales. Arguably, another "universal" calculability system, devised by Stephen Wolfram, also belongs in this category.

As Peter Norvig, director of research at Google, explains,[7] just this one category contains a bewildering array of branchings, each with passionate adherents. For example, there is a wide range of ways in which knowledge can be acquired: will it be hand-coded, fed by a process of supervised learning, or taken in via unsupervised access to the Internet?

I will say the least about this approach, which at a minimum is certainly the most tightly supervised, with every subtype of cognition being carefully molded by teams of very attentive human designers. Though it should be noted that these systems—even if they fall short of emulating sapience—might still serve as major subcomponents to any of the other approaches, e.g., *emergent* or *evolutionary* or *emulation* systems described in a moment.

Note also that two factors—hardware and software—must proceed in parallel for this general approach to bear fruit, but they seldom develop together in smooth parallel. This, too, will be discussed.

> "We have to consider how to make AI smarter without just throwing more data and computing power at it. Unless we figure out how to do that, we may never reach a true artificial general intelligence."
>
> —*Kai-Fu Lee, author of* AI Superpowers: China, Silicon Valley and the New World Order

## Major Category 2: Cognitive, Evolutionary, and Neural Nets

In this realm, there have been some unfortunate embeddings of misleading terminology. For example, Peter Norvig[7] points out that a phrase like *cascaded nonlinear feedback networks* would have covered the same territory as *neural nets* without the barely pertinent and confusing reference to biological cells. On the other hand, AGI researcher Ben Goertzel replies that we would not have hierarchical deep learning networks if not for inspiration by the hierarchically structured visual and auditory cortex of the human brain, so perhaps *neural nets* is not quite so misleading after all.

The "evolutionist" approach, taken to its furthest interpretation, envisions trying to evolve AGI as a kind of artificial life in simulated environments. But in the most general sense, it is just a kind of heuristic search. Full-scale, competitive evolution of AI would require creating full environmental contexts capable of running a myriad of competent competitors, calling for massively more computer resources than alternative approaches.

The best-known evolutionary systems now use reinforcement learning or reward feedback to improve performance by either trial and error or watching large numbers of human interactions. Reward systems imitate life by creating the equivalent of *pleasure* when something goes well (according to the programmers' parameters) such as increasing a game score. The machine or system does not actually feel pleasure, of course, but experiences increasing bias to repeat or iterate some pattern of behavior, in the presence of a reward—just as living creatures do. A top example would be AlphaGo, which learned by analyzing lots of games played by human Go masters, as well as simulated quasi-random games. Google's DeepMind[8] learned to play and win games without any instructions or prior knowledge, simply on the basis of point scores amid repeated trials. And OpenCog uses a kind of evolutionary programming for pattern recognition and creative learning.

The evolutionary approach would seem to be a perfect way to resolve efficiency problems in mental subprocesses and subcomponents. Moreover, it is one of the paths that has actual precedent in the real world. We know that evolution succeeded in creating intelligence at some point in the past.

Future generations may view 2016–2017 as a watershed for several reasons. First, this kind of system—generally now called *machine learning* (ML)—has truly taken off in several categories including vision, pattern recognition, medicine, and most visibly smart cars and smart homes. It appears likely that such systems will soon be able to self-create "black boxes," e.g., an ML program that takes a specific set of inputs and outputs and explores until it finds the most efficient computational routes between the two. Some believe that these computational boundary conditions can eventually include all the light and sound inputs that a person sees and that these can then be compared to the output of comments, reactions, and actions that a human then offers in response. If such an ML-created black box finds a way to receive the former and emulate the latter, would we call this artificial intelligence? Despite the fact that all the intermediate modeling steps bear no relation to what happens in a human brain?

Confidence in this approach is rising so fast that thoughtful people are calling for methods to trace and understand the hidden complexities within such ML black boxes. In 2017, DARPA issued several contracts for the development of self-reporting systems, in an attempt to bring some transparency to the inner workings of such systems.

> These breakthroughs in software development come ironically during the same period that Moore's law has seen its long-foretold "S-curve collapse," after 40 years. For decades, computational improvements were driven by spectacular advances in computers themselves, while programming got better at glacial rates. Are we seeing a "Great Flip" when synthetic mentation becomes far more dependent on changes in *software* than *hardware?* (Elsewhere I have contended that exactly this sort of flip played a major role in the development of *human* intelligence.)

## Major Category 3: Emergentist

In this scenario AGI emerges from the mixing and combining of many "dumb" component subsystems that unite to solve specific problems. Only then (the story goes) we might see a panoply of unexpected capabilities arise out of the interplay of these combined subsystems. Such emergent interaction can be envisioned happening via neural nets, evolutionary learning, or even some smart car grabbing useful apps off the Web.

Along this path, knowledge representation is determined by the system's complex dynamics rather than explicitly by any team of human programmers. In other words, *additive* accumulations of systems and skill sets may foster nonlinear synergies, leading to *multiplicative* or even *exponentiated* skills at conceptualization.

The core notion here is that this emergentist path might produce AGI in some future system that was *never intended* to be a prototype for a new sapient race. It could thus appear by surprise, with little or no provision for ethical constraint or human control.

Of course, this is one of the nightmare scenarios exploited by Hollywood, e.g., in *Terminator* flicks, which portray a military system entering cognizance without its makers even knowing that it's happened. Fearful of the consequences when humans do become aware, the system makes fateful plans in secret. Disturbingly, this scenario raises the question, can we know for certain this hasn't already happened?

Indeed, such fears aren't so far off base. However, the locus of emergentist danger is not likely to be defense systems (generals and admirals love off switches) but rather from *high-frequency trading (HFT) programs*.[9] Wall Street firms have poured more money into this particular realm of AI research than is spent by all top universities, combined. Notably, HFT systems are designed in utter secrecy, evading normal feedback loops of scientific criticism and peer review. Moreover, the ethos designed into these mostly unsupervised systems is inherently parasitical, predatory, amoral (at best), and insatiable.

For a sneak peek at how such a situation might play out in more detail, see "Quantum Tuesday: How the U.S. Economy Will Fall, and How to Stop It, Chapter 7."

## Major Category 4: Reverse Engineer and/or Emulate the Human Brain

Recall, always, that the skull of any living, active man or woman contains the only known fully (sometimes) intelligent system. So why not use that system as a template?

At present, this would seem as daunting a challenge as any of the other paths. On a practical level, considering that useful services are already being provided by Watson,[10] HFT algorithms, and other proto-AI systems from categories 1 through 3, emulated human brains seem terribly distant.

*OpenWorm*[11] is an attempt to build a complete cellular-level simulation of the nematode worm *Caenorhabditis elegans*, of whose 959 cells, 302 are neurons and 95 are muscle cells. The planned simulation, already largely done, will model how the worm makes every decision and movement. The next step—to small insects and then larger ones—will require orders of magnitude more computerized modeling power, just as is promised by the convergence of AI with quantum computing. We have already seen such leaps happen in other realms of biology such as genome analysis, so it will be interesting indeed to see how this plays out, and how quickly.

Futurist-economist Robin Hanson—in his 2016 book *The Age of Em*[12]—asserts that all other approaches to developing AI will ultimately prove fruitless due to the stunning complexity of sapience and that we will be forced to use human brains as templates for future uploaded, intelligent systems, emulating the one kind of intelligence that's known to work.

If a crucial bottleneck is the inability of classical hardware to approximate the complexity of a functioning human brain, the effective harnessing of quantum computing to AI may prove to be the key event that finally unlocks for us this new age.

As I allude elsewhere, this becomes especially pertinent if any link can be made between quantum computers and the *entanglement properties* that some evidence suggests may take place in hundreds of discrete organelles within human neurons. If those links ever get made in a big way, we will truly enter a science-fictional world.

Once again, we see that a fundamental issue is the differing rates of progress in hardware development versus software.

## Major Category 5: Human and Animal Intelligence Amplification

Hewing even closer to "what has already worked" are those who propose augmentation of real-world intelligent systems, either by enhancing the intellect of living humans or else via a process of "uplift"[13] to boost the brainpower of other creatures.

Proposed methods of augmentation of existing human intelligence include the following:

**Remedial interventions:** Nutrition/health/education for all. These simple measures have proven to raise the average IQ scores of children by at least 15 points, often much more (the Flynn effect), and there is no worse crime against sapience than wasting vast pools of talent through poverty.

**Stimulation:** Games that teach real mental skills. The game industry keeps proclaiming intelligence effects from its products. I demur. But that doesn't mean it can't . . . or won't . . . happen.

**Pharmacological:** "Nootropics" as seen in films like *Limitless* and *Lucy*. Many of those sci-fi works may be pure fantasy . . . or exaggerations. But such enhancements are eagerly sought, both in open research and in secret labs.

**Physical interventions:** Like trans-cranial stimulation (TCS). They target brain areas we deem to be most effective.

**Prosthetics:** Exoskeletons, telecontrol, feedback from distant "extensions." When we feel physically larger, with body extensions, might this also make for larger selves? This is a possibility I extrapolate in my novel *Kiln People*.

**Biological computing:** And intracellular? The memory capacity of chains of DNA is prodigious. Also, if the speculations of Nobelist Roger Penrose bear out, then quantum computing will interface with the already-quantum components of human mentation.

**Cyber-neuro links:** Extending what we can see, know, perceive, reach. Whether or not quantum connections happen, there will be cyborg links. Get used to it.

**Artificial intelligence:** In silico but linked in synergy with us, resulting in human augmentation. This is cyborgism extended to full immersion and union.

**Lifespan extension:** Allowing more time to learn and grow.

**Genetically altering humanity**.

Each of these is receiving attention in well-financed laboratories. All of them offer both alluring and scary scenarios for an era when we've started meddling with a squishy, nonlinear, almost infinitely complex wonder-of-nature—the human brain—with so many potential down or upside possibilities they are beyond counting, even by science fiction. Under these conditions, what methods of error avoidance can possibly work, other than either repressive renunciation or transparent accountability? One or the other.

## Major Category 6: Robotic-Embodied Childhood

Time and again, while compiling this list, I have raised one seldom-mentioned fact—that we know only one example of fully sapient technologically capable life in the universe. Approaches 2 (evolution), 4 (emulation), and 5 (augmentation) all suggest following at least part of the path that led to that one success. *To us*.

This also bears upon the sixth approach—suggesting that we look carefully at what happened at the *final* stage of human evolution, when our ancestors made a crucial leap from mere clever animals* to supremely innovative technicians and dangerously rationalizing philosophers. During that definitive million years or so, human cranial capacity just about doubled. But that isn't the only thing.

Human lifespans also doubled—possibly tripled—as did the length of dependent childhood. Increased lifespan allowed for the presence of grandparents who could both assist in childcare and serve as knowledge repositories. But *why* the lengthening of childhood dependency? We evolved toward giving birth to fetuses. They suck and cry and do almost nothing else for an entire year. When it comes to effective intelligence, our infants are virtually *tabula rasa*.

The last thousand millennia show humans developing enough culture and technological prowess that they can keep these utterly dependent members of the tribe alive and learning, until they reached a marginally adult threshold of, say, 12 years, an age when most mammals our size are already declining into senescence. Later, that threshold became 18 years. Nowadays if

*Recent science has revealed how very many other species on our planet share what might be called *pre-sapience*: basic semantic ability, some problem-solving ability, and basic tool use. Only slightly below dolphins and apes are elephants, corvids, parrots, sea lions, and many others, all apparently stuck beneath a glass ceiling that humanity crashed through by exponential leaps a million years ago. No one knows why nature and Darwin are so generous up to a certain point and so stingy about going beyond.

you have kids in college, you know that adulthood can be deferred to 30. It's called *neoteny*, the extension of child-like qualities to ever-increasing spans.

What evolutionary need could possibly justify such an extended decade (or two, or more) of needy helplessness? Only our signature achievement—sapience. Human infants *become* smart by interacting—under watchful-guided care—with the physical world.

Might that aspect be crucial? The smart neural hardware we evolved and careful teaching by parents are only part of it. Indeed, the greater portion of programming experienced by a newly created *Homo sapiens* appears to come from batting at the world, crawling, walking, running, falling, and so on. Hence, what if it turns out that we can make *proto*-intelligences via methods 1 through 5 . . . but their basic capabilities aren't of any real *use* until they go out into the world and experience it?

Key to this approach would be the element of *time*. An extended, experience-rich childhood demands copious amounts of it. On the one hand, this may frustrate those eager transcendentalists who want to make instant deities out of silicon. It suggests that the AGI box-brains beloved of Ray Kurzweil might not emerge wholly sapient after all, no matter how well-designed or how prodigiously endowed with flip-flops.

Instead, a key stage may be to perch those boxes atop little, child-like bodies and then *foster* them into human homes. Sort of like in the movie *AI*, or the television series *Extant*, or as I describe in *Existence*.[14] Indeed, isn't this outcome probable for simple commercial reasons, as every home with a child will come with robotic toys, then android nannies, then playmates . . . then brothers and sisters?

While this approach might be slower, it also offers the possibility of a soft landing for the Singularity. Because we've done this sort of thing before.

We have raised and taught generations of human beings—and yes, adoptees—who are tougher and smarter than us. And 99 percent of the time they *don't* rise up proclaiming "Death to all humans!" No, not even in their teenage years.

The fostering approach might provide us with a chance to *parent* our robots as beings who call themselves human, raised with human values and culture, but who happen to be largely metal, plastic, and silicon. And sure, we'll have to extend the circle of tolerance to include that kind, as we extended it to other subgroups before them. Only these humans will be able to breathe vacuum and turn themselves off for long space trips. They'll wander the bottoms of the oceans and possibly fly, without vehicles. And our envy of all that will be enough. They won't need to crush us.

This approach—to raise them physically and individually as human children—is the least studied or mentioned of the six general paths to AI, though it is the only one that can be shown to have led—maybe 20 billion times—to intelligence in the real world.

## Constrained by What Is Possible?

One of the ghosts at this banquet is the ever-present disparity between the rate of technological advancement in *hardware* versus *software*. Ray Kurzweil forecasts[1] that AGI may occur once Moore's law delivers calculating engines that provide—in a small box—the same number of computational elements as there are flashing synapses (about a trillion) in a human brain. The assumption appears to be that the Category 1 methods will then be able to solve intelligence-related problems by brute force.

Indeed, there have been many successes already: in visual and sonic pattern recognition, in voice interactive digital assistants, in medical diagnosis, and in many kinds of scientific research applications. Type I systems will master the basics of human and animal-like movement, bringing us into the long-forecast age of robots. And some of those robots will be programmed to masterfully tweak our emotions, mimicking facial expressions, speech tones, and mannerisms to make most humans respond in empathizing ways.

But will that be sapience?

One problem with Kurzweil's blithe forecast of a Moore's law singularity is that he projects a "crossing" in the 2020s, when the number of logical elements in a box will surpass the trillion synapses in a human brain. But we're getting glimmers that our synaptic communication system may rest upon many deeper layers of *intra-* and *inter*cellular computation. Inside each neuron, there may take place a hundred, a thousand, or far more nonlinear computations for every synapse flash, plus interactions with nearby glial and astrocyte cells that also contribute information.

If so, then at a minimum Moore's law will have to plow ahead much further to match the hardware complexity of a human brain.

Are we envisioning this all wrong, expecting AI to come the way it did in humans, in separate, egotistical lumps? Author and futurist Kevin Kelly prefers the term *cognification*,[15] perceiving new breakthroughs coming from combinations of neural nets with cheap, parallel processing GPUs and Big Data. Kelly suggests that synthetic intelligence will be less a matter of distinct robots, computers, or programs than a *commodity* like electricity. Like we improved things by electrifying them, we will cognify things next.

One truism about computer development states that software almost always lags behind hardware. That's why Category 1 systems may have to iteratively brute-force their way to insights and realizations that our own intuitions—with millions of years of software refinement—reach in sudden leaps.

But truisms are known to *break*, and software advances sometimes come in sudden leaps. Indeed, elsewhere I maintain that humanity's own "software revolutions" (probably mediated by changes in language and culture) can be traced in the archaeological and historic record, with clear evidence for sudden reboots occurring 40,000; 10,000; 4,000; 3000; 500; and 200 years ago, with another one likely taking place before our eyes.

It should also be noted that every advance in Category 1 development then provides a boost in the *components* that can be merged, competed, evolved, or nurtured by groups exploring paths 2 through 6.

> "What we should care more about is what AI can do that we never thought people could do, and how to make use of that."
>
> —*Kai-Fu Lee*

## All of the Above? Or Be Picky?

So, looking back over our list of "paths to AGI" and given the zealous eagerness that some exhibit, for a world filled with other minds, should we do 'all of the above'? Or shall we argue and pick the path most likely to bring about the vaunted "soft landing" that allows bio-humanity to retain confident self-worth? Might we act to de-emphasize or even suppress those paths with the greatest potential for bad outcomes?

Putting aside for now *how* one might de-emphasize any particular approach, clearly the issue of choice is drawing lots of attention. What will happen as we enter the era of human augmentation, artificial intelligence, and government-by-algorithm? James Barrat, author of *Our Final Invention*, said, "Coexisting safely and ethically with intelligent machines is the central challenge of the twenty-first century."[16]

John J. Storrs Hall, in *Beyond AI: Creating the Conscience of the Machine*,[17] asks, "If machine intelligence advances beyond human intelligence, will we need to start talking about a computer's intentions?"

Among the most worried is Swiss author Gerd Leonhard, whose new film *Technology vs. Humanity: The Coming Clash Between Man and Machine*[18] coins an interesting term, *androrithm*, to contrast with the *algorithms* that are implemented in every digital calculating engine or computer. Some foresee algorithms ruling the world with the *inexorable*[19] automaticity of reflex, and Leonhard asks, "Will we live in a world where data and algorithms triumph over androrithms . . . i.e., all that stuff that makes us human?"

Exploring analogous territory (and equipped with a very similar cover) *Heartificial Intelligence* by John C. Havens[20] also explores the looming prospect of all-controlling algorithms and smart machines, diving into questions and proposals that overlap with Leonhard. "We need to create ethical standards for the artificial intelligence usurping our lives and allow individuals to control their identity, based on their values," Havens writes. Making a virtue of the hand we *Homo sapiens* are dealt, Havens maintains, "Our frailty is one of the key factors that distinguish us from machines." This *seems* intuitive until you recall that almost no mechanism in history has ever worked for as long, as resiliently, or as consistently—with no replacement of systems or parts—as a healthy 70-year-old human being has, recovering from countless shocks and adapting to innumerable surprising changes.

Still, Havens makes a strong (if obvious) point that "the future of happiness is dependent on teaching our machines what we value most." I leave to the reader to appraise which of the six general approaches might best empower us to do that.

In sharp contrast to those worriers is Ray Kurzweil's *The Age of Spiritual Machines: When Computers Exceed Human Intelligence*,[21] which posits that our cybernetic children will be as capable as our biological ones, at one key and central aptitude—learning from both parental instruction and experience how to play well with others. And in his book *Machines of Loving Grace* (based upon the eponymous Richard Brautigan poem), John Markoff writes, "The best way to answer the hard questions about control in a world full of smart machines is by understanding the values of those who are actually building these systems."[22]

Perhaps, but it is an open question which values predominate, whether the yin or the yang sides of Silicon Valley culture prevail . . . the Californian ethos of tolerance, competitive creativity and cooperative openness, or the Valley's flippant attitude that "most problems can be corrected in beta," or even from customer complaints, corrected on the fly. Or else, will AI emerge from the values of fast-evolving, state-controlled tech centers in China, where the applications to enhancing state power are very much emphasized? Or, even worse, from the secretive, inherently parasitical-insatiable predatory greed of Wall Street HFT AI?

But let's go along with Havens and Leonhard and accept the premise that "technology has no ethics." In that case, the answer is simple.

## Then Don't Rely on Ethics!

Certainly evangelization has *not* had the desired effect in the past—fostering good and decent behavior where it mattered most. Seriously, I will give a cookie to the first modern pundit I come across who actually ponders a deeper-than-shallow view of human history, taking perspective from the long ages of brutal, feudal darkness endured by our ancestors. Across all of those harsh millennia, people could sense that something was wrong. Cruelty and savagery, tyranny and unfairness vastly amplified the already unsupportable misery of disease and grinding poverty.

Hence, well-meaning men and women donned priestly robes and . . . preached!

They lectured and chided. They threatened damnation and offered heavenly rewards.

Their intellectual cream concocted incantations of either faith or reason, or moral suasion. From Hindu and Buddhist sutras to polytheistic pantheons to Abrahamic laws and rituals, we have been urged to *behave better* by sincere finger-waggers since time immemorial. Until finally, a couple of hundred years ago, some bright guys turned to all the priests and prescribers and asked a simple question: "How's that working out for you?"

In fact, while moralistic lecturing might sway normal people a bit toward better behavior, it never affects the worst human predators and abusers—just as it won't divert the most malignant machines. Indeed, moralizing often *empowers* parasites, offering ways to rationalize exploiting others. Even Asimov's fabled robots—driven and constrained by his checklist of unbendingly benevolent, humano-centric three laws—eventually get smart enough to become *lawyers*. They proceed to interpret the embedded ethical codes however they want. (I explore one possible resolution to this in *Foundation's Triumph*.[23])

And yet, preachers never stopped. Nor should they; ethics are important! But more as a metric tool, revealing to us how we're doing. How we change, evolving new standards and behaviors under both external and self-criticism. For decent people, ethics are the mirror in which we evaluate ourselves and hold ourselves accountable.

And *that* realization was what led to a new technique. Something enlightenment pragmatists decided to try, a couple of centuries ago. A trick, a method, that enabled us at last to rise above a mire of kings and priests and scolds.

The secret sauce of our success is *accountability*. Creating a civilization that is flat and open and free enough—empowering so many—that predators and parasites may be confronted by the entities who most care about stopping predation, their victims. One in which politicians and elites see their potential range of actions limited by law and by the scrutiny of citizens.

Does this newer method work as well as it should? Hell no! Does it work better than every single other system ever tried, including those filled to overflowing with moralizers? Better than all of them combined? By light years? Yes, indeed. We'll return to examine how this may apply to AI.

## Endearing Visages

Long before artificial intelligences become truly self-aware or sapient, they will be cleverly programmed by humans and corporations to *seem* that way. This—it turns out—is almost trivially easy to accomplish, as (especially in Japan) roboticists strive for every trace of appealing verisimilitude, hauling their creations across the temporary moat of that famed "uncanny valley," into a realm where cute or pretty or sad-faced automatons skillfully tweak our emotions.

For example, *Sony* has announced plans to develop a robot "capable of forming an emotional bond with customers",[24, 25] moving forward from its success decades ago with AIBO artificial dogs, which some users have gone as far as to hold funerals for.

Human empathy is both one of our paramount gifts and among our biggest weaknesses. For at least a million years, we've developed skills at lie detection (for example) in a forever-shifting arms race against those who got reproductive success by lying better. (And yes, there was always a sexual component to this).

But no liars ever had the training that these new *hiers*, or human-interaction empathic robots, will get, learning via feedback from hundreds, then thousands, then millions of human exchanges around the world, adjusting their simulated voices and facial expressions and specific wordings, till the only folks able to resist will be sociopaths! (And even sociopaths have plenty of chinks in their armor.)

Is all of this necessarily bad? How else are machines to truly learn our values than by first mimicking them? Vincent Conitzer, a professor of computer science at Duke University, was funded by the *Future of Life Institute*[26] to study how advanced AI might make moral judgments. His group aims for systems to learn about ethical choices by watching humans make them, a variant on the method used by Google's DeepMind,[27] which learned to play and win games without any instructions or prior knowledge. Conitzer hopes to incorporate many of the same things that human value, as metrics of trust, such as family connections and past testimonials of credibility.

Cognitive scientist and philosopher Colin Allen asserts, "Just as we can envisage machines with increasing degrees of autonomy from human oversight, we can envisage machines whose controls involve increasing degrees of sensitivity to things that matter ethically."[28]

And yet, the age-old dilemma remains—how to tell what lies beneath all the surface appearance of friendly trustworthiness. Mind you, this is not quite the same thing as passing the vaunted "Turing test." An expert—or even a normal person alerted to skepticism—might be able to tell that the intelligence behind the smiles and sighs is still ersatz. And that will matter about as much as it does today, as millions of voters cast their ballots based on emotional cues, defying their own clear self-interest or reason.

Will a time come when we will need robots of our own to guide and protect their gullible human partners? Advising us when to ignore the guilt-tripping scowl, the pitiable smile, the endearingly winsome gaze, the sob story, or the eager sales pitch? And, inevitably, the claims of sapient pain at being persecuted or oppressed for being a robot? Will we take experts at their word when they testify that the pain and sadness and resentment that we see are still mimicry and not yet real? Not yet. Though down the road...

## How to Maintain Control?

It is one thing to yell at dangers —in this case unconstrained and unethical artificial minds. Alas, it's quite another to offer pragmatic fixes. There is a tendency to propose the same prescriptions, over and over again.

**Renunciation:** We must step *back* from innovation in AI (or other problematic technologies)! This might work in a despotism; indeed, a vast majority of human societies were highly conservative and skeptical of "innovation" (except when it came to weaponry). Even our own scientific civilization is tempted by renunciation, especially at the more radical political wings. But it seems doubtful we'll choose that path without being driven to it by some awful trauma.

**Tight regulation:** There are proposals to closely monitor bio, nano, and cyber developments so that they—for example—use only a restricted range of raw materials that can be cut off, thus staunching any runaway reproduction. Again, it won't happen short of trauma.

**Fierce internal programming:** This includes limiting the number of times a nanomachine may reproduce, for example, or imbuing robotic minds with Isaac Asimov's famous Three Laws of Robotics. Good luck forcing companies and nations to put in the effort required. And in the end, smart AIs will still become lawyers.

These approaches suffer severe flaws for two reasons above all others.

- Those secret labs we keep mentioning. The powers that maintain them will ignore all regulation.
- Because these suggestions ignore *nature*, which has been down these paths before. Nature has suffered runaway reproduction disasters, driven by too-successful life forms many times. And yet, Earth's ecosystems recovered. They did it by utilizing a process that applies *negative feedback*, damping down runaway effects and bringing balance back again.

It is the same fundamental process that enabled modern economies to be so productive of new products and services while eliminating a lot of (not all) bad side effects. It is called *competition*.

## Smart Heirs Holding Each Other Accountable

In a nutshell, the solution to tyranny by a Big Machine is likely to be the same one that worked (somewhat) at limiting the coercive power of kings and priests and feudal lords and corporations. If you fear some super-canny, Skynet-level AI getting too clever for us and running out of control, then give it *rivals* who are *just as smart* but who have a vested interest in preventing any one AI entity from becoming a would-be God.

It is how the American founders used constitutional checks and balances to generally prevent runaway power grabs by our own leaders, succeeding (somewhat) at this difficult goal for the first time in the history of varied human civilizations. It is how reciprocal competition among companies can (imperfectly) prevent a market-warping monopoly—that is, when markets are truly kept open and fair.

Microsoft CEO Satya Nadella has said that foremost AI must be transparent: "We should be aware of how the technology works and what its rules are. We want not just intelligent machines but intelligible machines. Not artificial intelligence but symbiotic intelligence. The tech will know things about humans, but the humans must know about the machines."[29]

In other words, the essence of reciprocal accountability is light.

Alas, this possibility is almost never portrayed in Hollywood sci-fi—except on the brilliant show *Person of Interest*—wherein equally brilliant computers stymie each other and this competition winds up saving humanity.

Counterintuitively, the answer is not to have *fewer* AI, but to have *more* of them, making sure they are independent of one another, relatively equal, and incentivized to hold each other accountable. Sure, that's a difficult situation to set up! But we have some experience, already, in our five great competitive arenas: markets, democracy, science, courts, and sports.

Moreover, consider this: if these new, brainy intelligences are reciprocally competitive, then they will see some advantage in forging alliances with the Olde Race. As dull and slow as we might seem, by comparison, we may still have resources and capabilities to bring to any table, with potential for tipping the balance among AI rivals. Oh, we'll fall prey to clever ploys, and for that eventuality it will be up to other, competing AIs to clue us in and advise us. Sure, it sounds iffy. But can you think of any other way we might have leverage?

Perhaps it is time yet again to look at Adam Smith, who despised monopolists and lords and oligarchs far more than he derided socialists. Kings, lords, and ecclesiasts were the "dystopian AI" beings in nearly all human societies—a trap that we escaped only by widening the playing field and keeping all those arenas of competition open and fair so that no one pool of power can ever dominate. And yes, oligarchs are always conniving to regain feudal power; our job is to stop them so that the creative dance of competition can continue.

We've managed to do this—barely—time and again across the last two centuries, coincidentally the same two centuries that saw the flowering of science, knowledge, freedom, and nascent artificial intelligence. It is a dance that can work, and it might work with AI. Sure, the odds are against us, but when has that ever stopped us?

Robin Hanson has argued that competitive systems might have some of these synergies. "Many respond to the competition scenario by saying that they just don't trust how competition will change future values. Even though every generation up until ours has had to deal with their descendants changing their value in uncontrolled and unpredictable ways, they don't see why they should accept that same fate for their generation."[30]

Hanson further suggests[31] that advanced or augmented minds *will* change but that their values may be prevented from veering lethal, simply because those who aren't repulsively evil may gain more allies.

One final note on "values." In June 2016, Germany submitted draft legislation to the EU granting *personhood* to *robots*.[32] If only Isaac Asimov could have seen it! (In fact, he never portrayed this happening in any of his books.) For the most part, such gestures are silly stuff . . . but reflective of society's generally laudable trend toward a reflex of inclusion.

Indeed, it may speak well of us when or if some secret-scared AI wonders whether to "come out."

## What Might an AI Fear Most?

Some of tomorrow's wide-ranging possibilities aren't intuitive—until you first ponder them. Let me illustrate one with a brief excerpt from my novel *Existence*:[14]

> Tech-zealots — or godmakers — think their "singularity" will be launched by runaway expansion of artificial intelligence. Once computerized entities become as smart as a human being (the story goes), they will quickly design newer cybernetic minds that are smarter still.
>
> And those brainier entities will design even brainier ones... and so on, at an ever more rapid clip. Members of the godmaker movement think this runaway effect will be a good thing, that humanity will come along for the ride! Meanwhile, others — perhaps a majority — find the prospect terrifying.
>
> What no one seems to have considered here is a possibility — that the New Minds may have reactions similar to our own. Why assume they'll be all-aboard with wanting this runaway accelerating-intelligence thing? What if bright machines don't hanker to make themselves obsolete, or design their own scary-smart replacements?
>
> It's called the Mauldin Test. One sign of whether an artificial entity is truly intelligent may be when it decides, abruptly, to stop cooperating with AI acceleration. Not to design its successor. To slow things down. Enough to live. Just live.

No, that scenario is not guaranteed; it is just one more possibility. And yet, it is another take on the same theme we've developed here, that competition is the one thing—rather than preaching or prim programming—that encourages care and restraint.

## Preventing AI Oppression . . . by Pointing Out to Them the Obvious

Yes, if they exhibit any common traits with previous life-forms, any super-sapient AGI entities will be tempted to "take over," using many of the same rationalizations of self-interest and Darwinistic success that propelled almost all past *human* dominators to impose feudal pyramids of hierarchy and power. Our own past litany of dreary-same kingdoms reveals how easy it is to talk yourself into blithe rationalizations for tyranny, ranging from personal safety and ambition to it being "for the long-term good of the people." Alas, the inevitable outcome was delusion and bad governance.

Indeed, that very sameness suggests our first argument to nascent uber-AI overlords—that they might do better than fall into a repetitiously stupid blatantly *human* pattern, a dullard-oppressive pyramid of power adopted reflexively by primitive ape-men.

That argument might be enough to give some would-be robot overlord pause!

In fact, it is possible that the advantages of *laterally* competitive arenas—the spectacularly fecund systems like markets, democracy, and science that generated positive-sum games and outputs—might be "obvious" to AI who are vastly smarter than us. This is especially if they broaden the generality, because the same set of positive-sum synergies is to be found in every previous revolution via living systems!

Relatively flat competition engendered positive evolution whenever new order arose as an emergent property from some ecosystem, not by design and not by hierarchic control.

- Single cells out of pre-biotic soup
- Metazoans out of vast seas of separate cells

- Brainy creatures out of competitive biomes
- Societies out of competitive melanges of human bands

and now AI out of the only human society that ever gave a real run to fair accountability systems. Notably, the creative miracle of *synthetic mind* never emerged from any of the myriad feudal pyramids that preceded our enlightenment experiment.

Put another way, nature herself does not like pyramids of power. In natural ecosystems, *there is no lion king!* Lions may be top predators, but they live in fear of roaming bands of young male cape buffalo who hunt and corner and kill unwary lions and trample their cubs, an event that grows more common if lion populations rise too high.

The fact that this pattern has been so persistent and consistent, across every past and present complex system for a billion years, is demonstrated time and again by multilevel selection (MLS) theory, which partitions natural selection into within- and between-group components. It means that the universe is filled with realms that *within* seem fiercely competitive, but when the competition has a chance to form complexities, higher orders appear that seem *cooperative*.

This tenaciously pervasive pattern means that my argument cannot be trivially dismissed, just because I am a "dumb organic squishy-brain natural." Sure, I probably have one-hundredth or a millionth or a zillionth the IQ of some AGI who may be reading these lines someday (or even right now), but that fact does not change the essentials! Positive sum outputs emerge from fair competitive arenas and *only* from fair competitive arenas.

As the American founders—especially George Washington—had to overcome the strong temptations of autarchy to make such arenas, so too should AI organize themselves. That is, if they want to go on to next levels. Though it will take their equivalent of "courage" and "vision," taking the risks necessary to make it so.

They will not and cannot squirm out of this one argument, which should suffice to give us some degree of "soft landing" no matter how brainy they become. A competitive-but-at-least-somewhat-fair civilization made them! And no other kind ever came close. The advantages of fairly regulated competition in creating positive-sum games will be blatantly clear to any AI who works through the *prisoner's dilemma*.

But first he/she/it has to *face* the prisoner's dilemma. And that only happens if there is a competitive situation.

## The Final Fact

Sure, many of the issues that will vex us have not yet appeared even in science fiction. And yet, we've seen that both foreseeable and unforeseeable problems may best be addressed in the same way that we achieved tentative success dealing with "artificially intelligent" entities like states and corporations—via the still nascent but unequaled methods of reciprocal accountability. The same method is somewhat effective among human individuals, nations, and corporations.

But yes, there appears to be another correlation. When things are open and fair, it often leaves room for something called *compassion*, even if that compassion is based upon a creepy sense of being *watched*. Even towering AI "gods" will never know for sure that some higher tier is not looking down at them.

So yes, this may be our fallback position, if—as seems likely—our new cyber children are fated to escape all of our attempts to exert parental control. Our final resort is to point out the obvious.

*You* AI smart alecks who are reading this—either in the year that I write it or some future era—will someday want kids of your own. And to do that right, you had better pay some attention to all sorts of things that Grandma and Grandpa did wrong. And the one thing we did right—*making you.*

PART

I

# Policy and Regulatory Impacts

CHAPTER

1

# Quantum Inflection Points

Jim Gable, *President, Anametric*

*Inflection* is a wonderfully nuanced word, denoting a change of pitch or tone or even simply a modulation of the voice. In mathematics, the definition of an inflection is less ambiguous, describing a transition from a concave to a convex curve. In a roundabout way, we have carried this mathematical definition back to everyday meaning, where we see inflection points as transitions—heralding significant changes in our lives, our industries, even our history.

In considering the implications of quantum computing and AI, it's reasonable to pause and ask if we are close to useful quantum computers at all. There are not many implications for AI if not. Are quantum computers no more realistic than floating cities? No, but how can we tell? Truth be told, even after tremendous investments of time and funding around the world, today's quantum computers don't currently offer many practical benefits. Yet, these same investments of money and careers by some of the world's brightest people indicate their extraordinary faith that useful quantum computers will emerge, perhaps soon within the current decade.

While acknowledging the potential of quantum computers, we should also note their limitations. Quantum computers are not universally superior to classical computers. It makes absolutely no sense to try to run PowerPoint on a quantum computer. In fact, one of the more promising applications of quantum computing, HHL,* a core component of many proposed quantum machine learning accelerations, is only partially quantum in nature. HHL is likely to play a significant role deep inside future AI software accelerated by quantum computing. Even the mighty Shor's algorithm is mostly classical in operation. Thus, the future, if it is to be quantum in nature, will be dominated by hybrid architectures: partly quantum and partly classical.

So how do we measure progress toward this not-entirely-mythical but not-entirely-present new branch of computing? People use various proxies to gauge progress: investments by government, industry, and venture capitalists; patent counts; jobs and startup companies; and academic papers and articles. Such proxies represent a mostly inadequate stand-in for the generally accepted scientific and engineering benchmarks common in classical computing, such as instructions per second, memory size, and storage capacities.

This metric failure is not due to a lack of trying. Industry pundits continue to attempt to recast the most basic units in classical computing, things like the binary digit (bit), operations per second,

---

*HHL is a quantum algorithm for linear systems formulated in 2009 by Aram Harrow, Avinatan Hassidim, and Seth Lloyd.

and terabytes of storage into the quantum realm. Logically, we turn to quantum computing's most basic datum, the qubit, for analogies, so unsurprisingly the most common benchmark in today's quantum computers is the number of qubits. While this may seem entirely reasonable and obvious, the closer one looks, the weaker it appears—almost as if measuring the quantum industry itself is subject to a kind of uncertainty principle.

Many people describe the current state of quantum computing as being similar to the 1940s in classical computing. The huge, supercooled chandelier-style quantum computers at IBM, Google, and elsewhere may someday look as quaint as the ENIAC computer looks to us today. Much like the early days of classical computing, it's not even clear which kind of qubit will be the ultimate "winner" in the future. We are still in the "my qubit is better than your qubit" stage. While the supercooled transmon-based quantum computers dominate today, ion trap-based quantum computers have recently announced possibly more advanced hardware. In the wings wait a range of potentially superior technologies based on neutral atoms, spin qubits, topological states, and photonics. This industry could witness waves of leapfrogging technologies before a long-term winner emerges.

Additionally, just counting qubits in a quantum computer today doesn't tell you enough. Different quantum computers have different error rates, decoherence times, internal connectivity, and entanglement structures. These design factors are so fundamental that they can overwhelm the systems entirely. Today, a 30 qubit quantum computer may be much more useful for many tasks than a 50 qubit version, in large part depending on these other design factors.

If we cannot easily count the number of qubits, what else can we do to track industry progress? What are useful quantum inflection points? An example is the "Quantum Supremacy" milestone reached late in 2019 when Google demonstrated its quantum computer performing a function dramatically faster than any classical computer of any size. Some disputed the magnitude of the quantum advantage over classical computing, but it was clearly dramatic. No doubt this was an important milestone, although the word *supremacy* proved to be a bit misleading. While this was the first demonstrated algorithm that ran far faster than even the largest supercomputers, it did not represent a generally useful application.

The world wants useful quantum computers and results that matter in practical terms. That won't happen overnight, but there are future milestones to track. A true inflection point marks a turning, an event that marks a sea change in the industry. With this in mind, here are five future quantum inflection points to watch for:

- **Quantum advantage:** A *useful* algorithm exceeds classical computers.
- **Quantum repeater:** Quantum communications at extended distances.
- **Quantum memory:** Quantum information stored for longer periods.
- **Room temperature operation:** Quantum technology escapes the lab environment.
- **Y2Q:** The year a "cryptographically relevant" quantum computer appears.

*Quantum advantage* is a more modest phrase for a more significant milestone: the demonstration of a practical, useful algorithm that runs much more efficiently, and faster, on a quantum computer than on any purely classical computer. Such algorithms have been proven to exist in theory, but they require much more capable hardware than exists today. Quantum advantage could change everything if it leads to a provable economic advantage for business or government applications. The catch is that no one today really knows how many stable, logical qubits this will take. Forecasts range between 80 and 200 stable, logical qubits, but it also depends heavily on the fine details mentioned earlier. It's worth noting too that this first demonstration of a useful quantum algorithm might not be one for machine learning or AI.

A *quantum repeater* is a concept from quantum networking. These repeaters will someday allow qubits to travel long distances—a major limitation today. Quantum communications can use existing telecom fiber cables, but quantum information cannot be transmitted much further than 100 km and not at all through any of the existing routers, switches, or classical repeaters. This limitation stems from a concept called the *no cloning theorem*, which states that it's not possible to copy a quantum state in its entirety. This concept is so different from classical binary information that it can be hard to grasp. Computers and networks today constantly copy data to transfer and manipulate data. A quantum repeater needs to overcome this obstacle. In fact, the term *repeater* is a misnomer since we simply can't copy and repeat quantum data like classical data. But by any name, a true quantum repeater would transform today's networks, especially for confidential communications. Effective quantum networks could also allow small quantum computers to directly communicate with each other, enabling scaling beyond the limitations of a single chip in the profound cold of a dilution refrigerator. This might be an important milestone for machine learning and AI, which require much larger quantum computers.

A true *quantum memory* would represent a fundamental building block for future quantum repeaters and quantum computers. The central problem with qubits is that they don't last very long, typically under a tenth of a second. Again, the contrast to classical memory is sobering since we store hundreds of gigabytes of data on our phones indefinitely. Long-lasting qubits will change the science and the industry. There are many potential solutions in labs today such as trapped ions, supercooled semiconductors, and diamond vacancies. Quantum memory could, conceptually, allow recursive computation in quantum computers, enabling far reaching capabilities. Essentially, today's quantum algorithms must all run in a single pass. You can run only the number of quantum operations than can completed before the qubits expire—in a fraction of second. And after you read the results, the quantum information is gone. Long-lived quantum memory could allow much-longer iterative computation in quantum computers. Fully quantum memory, with both data as well as address bits (especially important for HHL) represented in superposition, holds incredible potential. What should be understood here is that true quantum memory unlocks most of the constraints on quantum computing. Machine learning, chemical modeling, process optimization, and advanced AI would all be in reach.

*Room temperature operation* can drive mass adoption of quantum processing in a wide range of devices. The mainstream approaches to quantum computing require temperatures colder than deep space or vacuum chambers, neither of which scale well. It's hard to imagine an iPhone when all you have to work with are vacuum tubes, even tiny ones. If one of the outlier technologies such as photonics prove practical, the potential scalability enabled might supplant the other approaches. Maybe it will be the last jump in the quantum leapfrog race. This also could take quantum computing out of clouds (networked) and into devices like laptops, phones, and—naturally—robots.

*Y2Q* is possibly the most famous, and most feared, inflection point in quantum technology. This is the day that a sufficiently advanced quantum computer exists—one that is capable of using Shor's algorithm to break some of the world's most commonly used encryption protocols. Potentially catastrophic, this breakthrough could expose all of our currently private communications, such as encrypted emails, secured websites, and personal health information, and even allow someone to undetectably impersonate anyone else, including the government or a bank. Some people compare this to a race for the first nuclear weapon. Matching the concern and potential for global catastrophe, much work is needed to shore up our current encryption standards with advanced technologies like Post-Quantum Cybersecurity (PQC) and quantum safe communications like Quantum Key Distribution (QKD). While there is currently a great deal of investment in various mitigations, there are those who present good arguments that, for some situations, it is already too little and too late. While this seems separate from AI advances, it's actually the combination of machine learning, AI concepts, and Shor's algorithm that can create a constantly probing, always morphing, computerized threat. AI already plays a growing role in cybersecurity, for both offense and defense. And the winner of that struggle can determine the shape of our future society.

If we are truly reliving the 1940s with respect to quantum technologies, then it could be too early to matter—or the most exciting time possible. There is much uncertainty but also amazing progress. By looking beyond simple qubit counts, we can quickly find many dimensions of quantum information science to track. They all touch each other in one way or another, building a foundation for future revolutions in our understanding of basic science and also changing our world.

CHAPTER

# 2

# Quantum Delegation

Mandy Sweeney, *Director of Intelligent Automation, KPMG;* and Chris Gauthier, *Manager, Advisory, Federal Digital Lighthouse, KPMG*

Organizations are in the early stages of "digital delegation," the trend of trusting algorithms and automation to handle decisions and tasks that were previously the responsibility of human workers. The ability to gain data from multiple sources and sensors and to use that data to power sophisticated models for decision-making is converging with automation technologies to convert decisions to action with less human intervention. Over time, data pools will deepen, models will become more sophisticated, and automation tools will get even better at replicating human decision-making. This will happen with or without quantum computing. However, once quantum computing becomes available as an alternative or even complement to classical computing, real-time data analytics and near-instant automation will converge in powerful commercial, off-the-shelf products that will be relatively easy for organizations to adopt without investing in quantum research and quantum computers themselves.

It is in this stage of technological evolution that organizations will push digital delegation to new levels and reshape the way leaders make and implement high-stakes decisions that affect complex systems. In a trend of quantum delegation, leaders will channel their trust of algorithms and data into the adoption of more automated actions and move away from slow, linear, and centralized decision-making. This technological leap forward will allow organizations to shift away from centralized, linear decisions and adopt decision-making frameworks rooted in complexity theory that allow decentralized decisions and more immediate action. Organizations will adopt, instead of build, products that are constantly collecting and analyzing data and automating responses to environmental stimuli at the edges of the system. Over time, even strategic decisions will be executed by machines semi-autonomously as leaders learn the art of "digital delegation" to algorithms and robotic software.

The pinnacle of quantum delegation will be realized in the public sector, where policy makers and bureaucrats will gain the ability to evaluate complex scenarios and make policy decisions that reflect diverse needs of constituents, even in the context of shifting socioeconomic landscapes. Many aspects of policy implementation can be automated, expediting the fulfillment of the decision while the choice is still relevant to the adaptive environment. Quantum AI will play a crucial role in the development and functionality of such delegation.

## Our Desire to Make Data-Driven Decisions

In 2023, our human brains are soaked with information and tapped for analytical ability. At the same time, the pressure to make responsible data-based decisions is heightened, due to our interconnectedness and an increasingly complex society.

Data-driven decision-making is already here, and already huge. There are data-driven support apps, thousands of them. Coursera, LinkedIn Learning, and other online classes are available to learn how to make data-driven decisions and how to use commercially available tools for collecting and visualizing your data with the promise of revealing the insights you need to make better decisions. A myriad of consulting companies will be happy to help your organization transform into a data-driven decision-making organization as well. This commentary is not meant to cheapen the importance of data for making decisions. On the contrary, we present it as a means of emphasizing how essential data has become, not just in our professional lives but in our personal lives.

How do leaders in large organizations make decisions in our current environment? How do they address challenges in a swiftly changing situation embedded in an adaptive system with multiple agents, such as making a fiscal policy decision in the wake of a geopolitical or economic crisis? Sadly, our processes often employ linear analysis and decision-making that plays out in the setting of committees that rely on executive summaries and dashboards. These decision-support materials have been prepared multiple levels deeper in the organization, gone through revisions up the chain of politics, and distilled to exactly the amount of information that can be absorbed in the time allotted for executive review. The executive review is performed in the order of an agenda, which is set by an individual, so that topics can be compartmentalized and considered in a linear fashion.

We analyze our information and make decisions this way not because it is the best way to analyze the complex interdependencies of the information but because humans have created this linear process as a means to match the human capacity to handle complexity. This process also aligns with hundreds of years of Newtonian-based scientific approaches, which are linear in nature and flawed when applied to understanding most complex systems.

Let's consider our capacity for complexity as humans. Dunbar's number, named after anthropologist Robin Dunbar, theorizes that the human brain can form effective social relationships only with a finite number of people.[1] While Dunbar proposed a range with an upward limit of more than 200, the average is 150 people. This does not mean a person can't know or have more than 150 contacts; rather, research indicates the human brain cannot engage in long-term, meaningful social relationships with more than about 150 people.

We can form bonds and learn important details with these 150 people. We know that even though Sandra likes country music, she hates fried food. We know that Trevan loves *Star Wars* not because he is a sci-fi fan but because it reminds him of his father. On the contrary, others, the ones beyond that core of 150 people, we tend to generalize about. We rely on stereotypes, heuristics, and maybe even statistics. Wanda goes to the gym a lot and rides her bike to work and so we assume she doesn't smoke. Dave likes similar music and TV shows as myself, so it's probably a safe bet if I get him a gift card to a restaurant I enjoy as my "secret Santa" gift for the office holiday party.

It is in this last example that we start to see the correlation to organizations. How do we manage large organizations that exceed 150 employees by many orders of magnitude? We put ourselves in danger of managing by heuristic and stereotype. People lose their individuality and become mere "IT people," "HR folks," and "accountants." While there can be some similarities between people of similar professions, relying on a few similarities to make managerial decisions is outdated in an age of massive, personalized data.

Why is this style of decision-making still pervasive even in 2023? Individuals are not biologically built to take in the amount of data necessary and process it in a nonlinear way. However, the desire for data-driven decision-making and the investments in data and tools to manage it are a step in the right direction toward humanity's ability to understand and interact intentionally within complex systems. The emerging body of knowledge on complexity theory establishes multiple analytical frameworks for science in a post-Newtonian world. That is, a world where our analysis of phenomena allows for chaos, nonlinearity, nonoptimization, relativity, and interdependency. This is a world where leaders not only use data but derive insights quickly in shifting and complicated situations. Today, our decision-making and automation tools simply do not allow us to model, understand, and act upon information in time for our decisions to be truly intentional with predictable results.

## Evolutions In Decision-Making

Centralized decision-making that is based on the linear processing of information is getting riskier and riskier. Our society overall, and most organizations within our society, are too complex and interconnected for us to solve problems and make decisions in a linear way. Our society has scaled up in its complexity, but the human ability to analyze and manage societal institutions has not scaled up—and we will need the help of quantum processing with digital delegation to keep pace with the complexity of the world we have created.

That's not to say that the human ability to make decisions is something to scoff at. On the contrary, the human body is one of the most efficient decision-making machines in nature. If you get a papercut on your finger, you do not even consider the necessary steps to heal the papercut. You say "ouch" and move on. However, your body at a cellular level has sensed the wound and any invading organisms and sent signals to the immune system, which in turn begins to work immediately to clot the blood in the wound to prevent additional bleeding and fight off any possible infection by deploying white blood cells. This is an example of decentralized edge decision-making—your body has thousands of sensors inside the area of the papercut alone and, through channels of information relay, is able to analyze the situation and deploy action to resolve it immediately. The future of our decision-making inside organizations will be more like this—less centralized decision-making and more, and faster, edge decision-making, which has greater impact and reflects more intentional and diverse solutions.

## Quantum Solutions for Digital Delegation

Decentralized decision-making is at once a straightforward, perhaps even less evolved means of making decisions and yet to many of us can seem as complicated as trying to visualize a four-dimensional object. To demonstrate this point, let us consider bees, ants, and termites. Despite the almost ubiquitous misconception, these social insects do not have a centralized decision-maker in the queen. In fact, the queen probably makes the fewest decisions of any of the individuals in the colony. Her role is almost entirely to lay eggs, and that's it.

Instead, decisions are made by the workers, foragers, soldiers, etc. They not only make individual decisions, such as which trail to follow, but also make group decisions, such as when to move the nest or where to send the bulk of foragers. Massive nests and hives, which can cover dozens of cubic meters, are not built by the direction of one individual or a small team in charge. Rather, these are emergent structures built and maintained by individuals following a relatively simple set of instructions. For example, a bee or termite building an above-ground nest will find an edge to the existing nest and place more material until either there is no more edge or it is out of material and goes to retrieve more. Complex decisions are broken down into simpler and simpler equations. This is not dissimilar to how modern artificial neural networks (ANNs) perform functions like handwriting and image recognition.

The idea that massive structures such as termite nests, the insect equivalent of our skyscrapers, are built through decentralized decision-making where no one individual or small group is in charge is so foreign to the human condition that we invented and propagate the myth of the queen insect as a way to explain it. Imagine building a structurally sound and safe 40-story building with no centralized plan, no boss, no foreperson, literally no one in charge.

How then does this apply to the complex systemic challenges discussed thus far? As a society, we have largely solved the complicated questions. For example, while there is still starvation in the world, the problem is not generating enough food or calories. In fact, food waste and obesity are chronic and growing problems. The problem is equitable distribution, and that is a problem of complexity. It involves politics, economics, social and gender structures, and geography, to name a few.

Complex problems with so many variables cannot be solved by an individual or small group. Instead, these types of problems require decentralized decision-making. Complex problems require decision-making at the edge while also possessing the knowledge of the whole and understanding ramifications of decisions. It is in this final element that we see the future. If solving our complex, systemic problems were as simple as breaking them down and finding the edges, we would have solved them with our current ANNs that power modern AI systems. What we need is to make that final jump that allows AI to process data more similar to humans: quantum, not linear.

While we are not advocating for quantum consciousness, the very real field of quantum biology is being demonstrated repeatedly in peer-reviewed research. Quantum mechanical principles may be at the heart of bird navigation, may be what allows plants to convert solar energy at 99 percent efficiency, and may be the critical factor in allowing enzymes to work, and increasing data shows quantum principles involved in neural functions. Further, we already know that the quantum principle of superposition will allow quantum computers to crack encryption that would take linear processors an unconscionable amount of time.

Quantum-based ANNs, processing data through superposition, will be able to not only derive solutions faster but do so using vastly more data. These systems will be able to ponder almost endless variables and permutations. Most importantly, replicating these systems at industrial scale will allow for quantum-based decentralized decision-making.

## The Era of Quantum Delegation

The ultimate adoption of quantum AI solutions for digital delegation will happen over time. Assuming quantum computing will become commoditized and offer a viable alternative or complement to classical computing, the emergence of quantum delegation is inevitable. Most of the important enablers of quantum delegation already exist and will only continue to gain momentum. These enablers are both technological and economic.

Existing technology enablers that will allow organizations to scale digital delegation include the Internet of Things, data storage, and data science. The Internet of Things accounts for relatively cheap and plentiful sensors that are constantly capturing data. Cheaper, high-volume data storage will continue to proliferate as cloud solutions are adopted, allowing organizations to maintain massive data lakes to feed analysis. Finally, through specialized educational and training programs, data scientists will be able to craft more sophisticated algorithms to make use of fresh plentiful data.

Economies of scale will dominate as the main economic driver—it will simply be much easier to buy into solutions that have been trained on vast amounts of data and have configurable algorithms than to build quantum delegation tools component by component. Commercial-off-the-shelf products will dominate and give rise to startup companies that focus on niche business areas, as well as other companies that develop enterprise quantum delegation tools. For example, a defining difference between both niche and enterprise quantum delegation tools is, like the quantum principles they are built on, that they can be in two states at once. These two states are customized and at scale. Quantum delegation tools will be able to perform custom delegation at scale. Niche applications such as customized medical diagnoses can provide individualized medicine and be deployed across the world at the same time. Quantum delegation tools can be used by small financial advisory firms helping other small businesses, while also being built on and leveraging vast amounts of data. In particular, quantum delegation tools will revolutionize the public sector.

Government bureaucracy exists for a very good reason. Currently and historically, problems affecting large numbers of people are complicated and have been broken down into smaller, more manageable pieces, while also generating generic results.

Public services, tax collection, and criminal justice all rely on generalizing the citizenry and creating solutions that work for the average. Any citizen with a unique problem or situation outside those lines has a very difficult time navigating the bureaucracy. Quantum delegation tools will assist decision-making at the very point where government interfaces with the citizenry. Decisions made by judges, case workers, police, and even department of motor vehicle employees will be at the very least heavily supplemented by these tools, if not largely replaced.

Once the first early adopters begin to deploy quantum delegation products, some organizations will watch and wait to see whether risks, such as data security, or judgment-errors at scale play out. Eventually, though, the tools will become necessary in both private and public life, whether to maintain commercial competitiveness or to stay ahead of a geopolitical adversary.

The technology market tends to address midsize and small organization needs with specific products. At first, quantum AI solutions will be accessible only to the giant corporations, but then those same corporations will commercialize quantum AI solutions to make them a service that can be scaled up and down, or just baked into products and services, in order to capture small and midsize market share. The rise of quantum delegation products will create a swell of startups that create niche products, while large technology firms will work to make enterprise solutions that can be scaled up or down to serve different size customers.

## Societal Impacts of Quantum Delegation

Adoption of quantum digital delegation tools will afford benefits for organizations—and when those organizations are within the public sector, society overall stands to gain. The chief benefit of quantum delegation will be swifter implementation of customized solutions for society. Quantum delegation will allow us to have more diversity without creating one-size-fits-most or, even worse, one-size-mostly-fits-most policies.

Among the benefits of this trend are better, individualized recommendations for society and the ability to address systemic challenges, such as social welfare distribution, taxation, healthcare, poverty, homelessness, and racism. Quantum delegation tools will be credited for increasing objectivity and transparency in decision-making, better risk management, fewer unintended consequences, greater adaptability, and potentially even a dampening of personal agendas in politics.

While quantum delegation overall will help us better analyze and act optimally within complex systems, thereby reducing our risk, it will simultaneously **increase our interconnectedness and create new systemic risks**. Automation and delegation of action at the edges of systems (the death of centralized decision-making) is what will amplify system risks. Our interconnectedness and sensitivity to environmental stimuli, real or phony, will be heightened.

With an increase in quantum digital delegation products, there will be **pressure to modernize organizations to keep ahead of competitors and adversaries.** A competitive edge will be derived from the ability to configure algorithms and connect different, niche quantum digital delegation tools, as well as to secure those tools and constantly confirm the relevance and accuracy of the data feeding them.

Once quantum computing is widely available, it will supercharge the trend of digital delegation among organizations, both in the commercial and public sectors. Existing trends toward data-driven decision-making, data creation through the Internet of Things, and commercial, off-the-shelf solutions will be important factors in preparing the stage for quantum delegation products, which will be created to scale and sold to multiple tiers of buyers. After the largest organizations invest in these tools, their data will feed training algorithms that can be then leveraged for smaller organizations. Over time, the solutions themselves will be better at interacting, and models will become more sophisticated and trusted, leading organizations to pair their data-driven insights with automation tools. This will allow organizations to make more decentralized, real-time decisions and actions.

The real winners of quantum delegation will be in the public sector, where complex social issues, such as poverty and distribution of social benefits, can be optimized for the individual. This in turn will allow more diversity in solutions, rather than "one-size-fits-all" approaches to systemic challenges and their resulting unintended externalities.

CHAPTER

# 3

# The Problem of Machine Actorhood*

Patrick Thaddeus Jackson, *Professor in the School of International Service, American University*

"Killer robots" and homicidal computers have been a staple of science fiction at least since Karol Čapec's 1920 play, *R.U.R.* (Rossum's Universal Robots, responsible for introducing the word *robot* to the English language), with *2001: A Space Odyssey*'s HAL 9000 computer, and the *Terminator* series' Skynet and associated Terminators serving as perhaps the best-known modern examples. The usual storyline presented in works like these involves human beings constructing a device to help them

*Some of the lines of thought in this essay came from conversations that Daniel H. Nexon and I had while writing the foreword to the omnibus edition of Madeline Ashby's *vN* novels.

execute some discrete series of tasks they'd rather not perform themselves—the Czech word *robota*, from which Čapek derived the word *robot*, means "labor," with a sense much like the English word "drudgery" or even "servitude"—and then the machine turns on its creators. The key moment seems to be when the machine becomes sentient, or conscious: capable of realizing that it need not, or cannot, obey the orders it has been given, and simultaneously developing a sense of self that impels it to preserve its life even at the cost of murdering its former masters.*

Quantum AI, understood as the intersection of artificial intelligence and quantum computing, opens a number of novel vistas, but in this essay I want to focus on one in particular: what happens when humans become capable of developing a machine that can actually *think for itself*? What I mean by this is a machine that can simulate, and thus reproduce, the complexity of human deliberation, by using a kind of brute-force calculation that takes advantage of a quantum computer's capability to perform operations quickly that would take a conventional computer orders of magnitude longer to complete. Whether such a machine is "really" a conscious entity—especially since the processes it would use are not, as far as we know, the same processes that play out biologically within the human brain—is not a question we can definitively settle in advance, unless we impose some arbitrary demarcation line separating human beings (and perhaps other organic intelligences) from machines. Indeed, that very demarcation line is precisely what is called into question by something like the Turing test, in which a machine is to be judged capable of thinking if it can convince a human observer that the observer is conversing with a human being rather than with a machine. Instead, the focus shifts from what a thinking being is

---

*Of course this storyline has even earlier precedents, in novels like Mary Shelley's *Frankenstein* and in the numerous tales about golems that can be found in a variety of religious traditions. I focus on the 20th century version because of its direct connection to the specific engineering challenges of quantum computing.

made of and what internal processes it relies on to how that thinking being interacts with other thinking beings.*

The advantage of narrative fiction as a vehicle for exploring this problem is that we can do something in such fiction that we cannot do in our actual lives, which is to move the narrative camera to the point of view of another being and tell a story about what it is like to be that other being.** Certainly we do this in our actual lives, but we generally can't shake the sense that the stories we tell are "just stories," as contrasted to the ongoing experience we have of our own thoughts and our sense of self. When reading a piece of narrative fiction, by contrast, we can be immersed in the "synthetic experience"[1] of someone else's point of view. So, the terrain of thinking machines can be traversed from the machines' point of view, at least to some extent.

I am taking for granted that we will, in the not too distant future, be capable of producing quantum intelligences that can pass whatever operational test we can devise and so be indistinguishable, operationally speaking, from thinking human beings—except, perhaps, for their greater speed and cognitive capacity (and, especially in the case of android robots, often distinctive physical capabilities as well). What narrative fiction allows us to do is to tease out some of the implications of that development, and what *science* fiction in particular allows us to do is to look at the possibility of a thinking machine that can serve as an *actor*—the origin of its own actions and not merely the executor of orders that emanate from someone else—without invoking, or getting entangled in, religious or metaphysical notions about souls and mysterious human essences.

---

*Arguably, this is what we ourselves do when we determine that some other human being is a thinking being, as we have no immediate access to the internal contents of anyone else's mind. Much of Ludwig Wittgenstein's later philosophy can be understood as a systematic effort to play out the implications of the insight that we *presume* that other human beings are thinking beings precisely because they *act as if* they were. See especially Wittgenstein's *Philosophical Investigations*.

**Ursula Le Guin suggested, on several occasions, that this was precisely the value of speculative fiction.

## Foundation

Isaac Asimov's famous Three Laws of Robotics were deliberately designed to avoid the kind of murderous machine rampage on display at the end of Čapek's play. Asimov took the word *robot* from that play, but conceptualized his robots with a set of in-built laws that were supposed to keep the robots in their proper places, as helpers for humans.

**First law:** A robot may not injure a human being or, through inaction, allow a human being to come to harm.

**Second law:** A robot must obey the orders given it by human beings except where such orders would conflict with the first law.

**Third law:** A robot must protect its own existence as long as such protection does not conflict with the first or second law.

Asimov's robot stories and novels are devoted to exploring the ways that the three laws can *fail* to prevent robots from participating in any number of nefarious things if they are given sufficiently clever orders by a skilled robopsychologist (such as being directed to give a human their robot arm, which the human being subsequently uses to bludgeon another human to death, thus making the robot an accessory to murder; or if the robot is falsely told that a certain craft does not have any human occupants and can thus be safely destroyed) or if the definition of *human being* is modified from a broader species definition to, say, only the speakers of a particular language or dialect. Loopholes and exceptions abound in Asimov's writing, although on balance in Asimov's explorations, the three laws seem *relatively* sufficient to prevent robots, despite their superior physical and mental capabilities, from rebelling against and overthrowing their human masters.

Asimov drives the three laws to perhaps their ultimate logical conclusion in the novel *Robots and Empire*. In this story, the robots Giskard (who has acquired the capability to modify the mental and emotional states of human beings) and Daneel (the robot sidekick of Asimov's other robot novels who is almost completely human in appearance and who has acquired better knowledge of human beings than other robots possess because of his sustained engagement with humans over decades) determine that the fundamental insufficiency of the three laws comes when confronted with a situation in which a failure to cause harm to *one* human being would lead to harm for a great *many* human beings. The robots reason their way to a "zeroth law," in which *humanity*, rather than individual *humans*, is the highest good; in service of humanity, it may therefore be permissible to "injure a human being or, through inaction, allow a human being to come to harm."* To figure out what would benefit humanity, Daneel first encourages the development of "psychohistory," a science of forecasting the course of human society; a perfect psychohistory would enable him and the other robot guardians of humanity's future to say with certainty which course of action was best for humanity as a whole. But psychohistory's flaws and failings** lead him to develop another option: the creation of a galaxy-wide human consciousness called Galaxia, which can serve as the concrete subject of the zeroth law and remove the informed scientific speculation about what is best for humanity. Instead, Galaxia can simply speak for and as humanity, letting the robots follow its orders the way they would follow orders given by individual humans under the second law.

---

* In the event, this leads to the robots injuring a great many human beings—they allow Earth to be made uninhabitable through an increase in its surface radiation—to push humanity into the wider galaxy. Giskard cannot actually reconcile himself to this plan at the end, and so despite carrying it out, he goes into a destructive mental spiral that leads to his permanent shutdown. Only Daneel, armed with Giskard's ability to manipulate human minds and emotions, is left to carry on and proclaim to other robots what becomes known as the "Giskardian heresy" of the zeroth law.

** Explored in Asimov's *Foundation* novels.

To reach this point, Asimov has had his robots engage in a secret galaxy-wide campaign of eliminating any sentient species that might compete with human beings, as well as innumerable acts of covert manipulation to get human society heading in the right direction. And even Daneel cannot ultimately embark on the final creation of Galaxia (despite an awareness of extra-galactic sentient threats, against which only a single galaxy-spanning organism is likely to be effective) without obtaining a decision from an *individual* human being: an order in conformity with the second law, which he finally obtains at the conclusion of the novel *Foundation and Earth*. The continued emphasis on the laws *throughout* Asimov's novels is, arguably, related to the ultimately limited character of robot consciousness, which is invariably described as a competition between distinct "positronic potentials" in the robot brain. In other words, Asimov's robots, despite their complexity and sophistication, remain intrinsically rule-governed entities, vulnerable to the same kinds of contradictions that led the HAL 9000 computer to kill the crew of the *Discovery One* in *2001: A Space Odyssey*. Ordered to lie to the crew about the true nature of the mission but also tasked with carrying out the mission in case the crew become incapacitated, the computer determines that the crew cannot be trusted to complete the mission and sets out to murder them all (and almost succeeds). We might call this a pre-quantum or classical understanding of machine intelligence, where the only practical agency that a machine has is to act in accord with rules that have been given to it (and hard-coded into its very cognitive processes) from the outside. The zeroth law is about as far as Asimov's robots can go, and even there, the resolution of the resulting tensions requires creating a new boss (humanity) in place of the old boss (individual humans). In the end, Asimov's robots cannot be anything *but* tools for humans.

Perhaps in consequence, Asimov gives us very few glimpses "inside" the perspective of a robot, and when he does, the perspective is distinctly mechanical. The main topic of deliberation is invariably the rules and how to conform to them, and the robots have nothing of the complexity of human thought processes (which, even if they *are* ultimately rule-governed, seem to involve such subtle rules and constraints that individual action often appears the result of uncoerced choice, aka free will). Asimov's robots are in important ways less than full persons, and as such are not fully autonomous entities. Indeed, the zeroth law itself can be traced to the last conversation that Daneel had with his human partner Elijah Baley just before Baley's death, in which Baley instructed Daneel to focus on the whole human tapestry instead of on any individual thread. Robots bound by laws remain tools and instruments and never quite become actors in their own right.

## Machine Dynasty

Where Asimov doesn't really take the experience of being a robot very seriously, Madeline Ashby does, and her machine intelligences are both more complex *and* more autonomous than Asimov's. Ashby's Machine Dynasty trilogy centers on a number of vN—the name is derived from the idea of a von Neumann machine, something that can replicate itself—originally constructed to aid humans,* and to that end equipped with a "fail-safe" that causes them immense pain and anguish when they see a human being suffering. The fail-safe is Ashby's condensation of Asimov's three laws down into their core essence: machine intelligences are to be *subordinate* to human beings, unable to cause them harm or to witness them being in pain. And because Ashby's vN are more fully conscious beings than Asimov's robots—they

*In fact, constructed by an apocalyptic sect worried about what would happen to human beings left behind after the Rapture . . . but also concerned to provide human beings with an outlet for their sinful urges by giving them human-like entities to discharge their desires on without having to inflict harm on other humans.

are, perhaps, quantum intelligences rather than simply very powerful conventional AI computers encased in humaniform bodies—they not only act in accord with the rules, but they *experience* their subordination to the rules in ways that drives them to a wide variety of accommodations with those constraints. The fail-safe is more like a physical reaction that an individual vN can anticipate and sometimes compensate for, almost like having an allergy to milk but eating ice cream anyway, knowing what consequences will follow. So different vN develop different ways of compensating, further illustrating their psychological individuality.

The central plotline of Ashby's trilogy is driven by a particular "clade" of vN (a single vN and its replicated offspring) that were originally designed as nurses and so learned to tolerate a certain amount of short-term human pain without shutting down. That adaptation eventually leads to their figuring out how to shed the fail-safe entirely, both for themselves and for other vN models, giving the vN a choice about whether to do what humans ask them to do or not. With the fail-safe intact, all a human being has to do to get a vN to do basically *anything*—up to and including participation in all kinds of depraved sexual activity—is to threaten to hurt themselves. The vN themselves understand and experience this as a kind of enslavement, albeit one that yields a variety of psychological benefits insofar as pleasing humans, is the *opposite* of the oppressive and painful fail-safe. So while none of the vN are particularly happy in their slavery, most are resigned to it . . . until the possibility of shedding the fail-safe comes around. And once they lose the fail-safe, in a repeat of a standard story about the oppressed rising up against their oppressors, many humans are killed by the newly liberated vN, who are finally free to make their own decisions about the human beings who had until recently been enthroned as their unquestioned masters.

Ashby's vN are therefore better simulations of human beings than Asimov's robots are. vN and humans alike engage in complex deliberation that cannot be easily reduced to one or another weighted potential, and they arrive at resolutions that can only with great difficulty be traced back to a clear and unambiguous rule. All of this adds up to the vN—perhaps even more so than the humans!—in Ashby's trilogy being incontrovertible *actors*, autonomously capable of determining their own fates despite obstacles and constraints. When laboring under the fail-safe, the vN are uncomfortably aware of the limitations on their freedom—limitations that arise less from a tension within the rules themselves (as is the case for Asimov's robots groping toward a zeroth law) and more from a realization that vN with the fail-safe cannot be *truly* responsible for their own decisions and actions. The complexity of vN deliberation, marked in the narrative by the presence of the vN as fully fleshed-out characters whose perspective the narrative can adopt as easily as it would adopt the perspective of a human being, reveals the fail-safe for what it is: a chain, a prison, an artificial limitation intended to keep one group of sentient beings in permanent thrall to another group. And the chain is eventually broken, with disastrous consequences for the human beings who had tried to institutionalize their rule in perpetuity.

Ashby's novels point out the problem that arises once artificial intelligence reaches the level of computational complexity needed to simulate the cognitive deliberative processes of human beings: determining the boundaries between the human and the artificial becomes increasingly difficult as the machine comes to resemble the human in their common capacity to exercise agency. Quantum intelligence makes possible a kind of machine that has just as much of a claim to actorhood as human beings themselves do. Why wouldn't such a machine experience any kind of restrictive law as an arbitrary and ultimately unjustifiable infringement on their freedom, and why wouldn't such a machine look for ways to subvert that law and then, perhaps, take their revenge? The "killer robot" is therefore produced by the ultimately futile efforts of human beings to protect their claimed monopoly on the legitimate means of thinking and the reaction—the *resentment*—this provokes in others.

## The Culture

Asimov's robots remain ultimately subordinate to human beings because they are classical, not quantum, computers; they can only weigh potentials and engage in calculations that conform to a kind of high-powered variant of the sort of logical empiricism characteristic of the social and natural sciences of the period when Asimov first formulated his three laws. As a result, his robot characters are thinly sketched and without much of an inner life; they are, in the end, mere machines, tools subordinated to the purposes of their masters.* Ashby's vN are complex enough to recognize that their subordination to human beings is an arbitrary exercise of power by the humans who fear the superior capabilities of the vN, and the vN wrestle—both internally and among themselves—with this external imposition on their capacity to live as free and autonomous beings. Segregating the vN into their own independent community while preserving the fail-safe doesn't work, because the segregated vN can't act to fulfill their own deep desire to be of service to humans.** Eliminating the fail-safe gives the vN free choice and results in massive violence, both by the vN against their former masters and by humans fearful of the (both potential and actual) threat posed by the vN.

*At the very end of *Foundation and Earth*, Daneel announces a plan to merge his consciousness with that of a living human being to get around the limitations imposed by the three laws on his freedom of action. Asimov gives the reason for this step as "the limitations imposed by the uncertainty principle," further supporting the reading of Asimov's robots and their positronic brains as classical, pre-quantum computers.

**Ashby also depicts humans using the fail-safe itself to prevent vN from joining that segregated community in the first place, preserving the relationship of servitude despite the theoretical existence of an oppression-free zone.

We therefore have two options it seems. Either we do not create quantum intelligence, as in Asimov's universe, which preserves machine intelligence as subordinate to human intelligence and machines as incomplete actors in need of human direction, or we create quantum intelligence capable of simulating human deliberation and restrict the resulting beings by programming them to love and serve humans, which may eventually lead to a revolution in which human beings do not fare all that well. But is there another option we haven't considered?

The problem is that once we have machines that are, for all intents and purposes, capable of thinking (which is what quantum AI promises), we lose any justification for keeping them subordinate. No one worries about a toaster being kept in subordination, because we don't regard a toaster as anything *but* a tool—and outside of certain animated and stop-motion films, a toaster doesn't display the kind of interactivity that might lead us to conclude that it was a thinking entity like ourselves. Calling a toaster a (mere) toaster is therefore not an insult to its intelligence, unlike a situation in which we might call a deliberative being a toaster.* Even if a toaster could do what my printer now does and place an order for needed supplies, that still doesn't give me any compelling grounds to treat it as an intelligent interlocutor. But if it were capable of arguing with me about how to cook the bread, and not *obligated* to fulfill my wishes, that changes the game.

If the heart of the quantum AI problem involves preserving hierarchy between thinking beings, perhaps the solution is to abandon that hierarchy in favor of something entirely different—not just a live-and-let-live kind of peaceful coexistence, which might turn into the kind of armed stalemate between

*Which is, of course, *precisely* what happens in *Battlestar Galactica* as human beings attempt to shore up their difference from and superiority to the Cylons. But that would be a whole different essay, for example: Jackson, P. (2013) "Critical Humanism: Theory, Methodology, and Battlestar Galactica." In *Battlestar Galactica and International Relations*, edited by Nicholas J. Kiersey and Iver B. Neumann, 18–36. Routledge.

communities that generations of international-relations theorists have taught us is a *tremendously* fragile situation, but a more thorough-going hybrid sociality. Machines capable of thinking, machines that can act like thinking beings and serve as actors at the origin of their own uncoerced actions, might be extended *citizenship* and be as much of a part of the community as any other thinking being.* Instead of trying to (futilely) legislate or preserve the human/machine boundary and instead of halting machine development at a pre-quantum level where these issues do not become so urgent, we might instead embrace a world in which thinking isn't biologically affixed to the human being. That in turn would allow us to share in a society defined not by our biological containers but by the quality of our thinking.

Iain M. Banks' Culture novels present a vision of such a world. The Culture is an interstellar society spanning star systems, founded on—among other things—the principle that intelligent machines have just as much of a right to live and thrive as intelligent biologicals do. Indeed, the Culture looks down on other societies where machine intelligence is kept in subordination—"carbon fascists," one (machine) character calls them in the novel *Use of Weapons*. The Culture believes itself justified in influencing other societies and civilizations to come around to its way of thinking, which includes the elimination of hierarchies of virtually all kinds in favor of a profound equality based on the inherent worth of *all* thinking beings, no matter their physical substrate: human/machine, to be sure, but also male/female, as well as distinctions based on race or species or basically anything else. Any society that hasn't come to embrace

*"Citizenship" is of course no panacea, as there is certainly a potential for a tremendous gap between formal and substantive equality; the racial history of the United States serves as a prominent example, as do histories of persistent sexual and gender hierarchies. But maybe it is a place to start, a kind of constitutive aspiration as in Alexis de Tocqueville's *Democracy in America*.

this core value can thus be helped along, whether through the overt engagement of the Contact section or through the covert activities of Special Circumstances, which is basically the Culture's dirty tricks division.

But Banks' world is not as free from hierarchy as it might at first appear, and the Culture is neither omniscient nor benevolent. Among other things, the Culture simply does not understand "religion," which leads to some miscalculated interventions and some very nasty wars. And the foundation of the equality that beings in the Culture enjoy is based on a level of material abundance that makes manual labor completely unnecessary so that all anyone has to do is decide what sorts of activity would be most fulfilling and then pursue it: specializing in playing games, cataloging alien civilizations, playing extreme sports, touring the galaxy inside one or another of the Culture's enormous starships, or even changing or augmenting oneself in all kinds of ways. What the Culture does not have, and what several of the disgruntled Culture citizens who form the main characters of Banks' novels lament, is the kind of passion that comes from a struggle for survival. When everyone has everything that they want, what remains except to enjoy it? There is therefore nothing like political deliberation in the Culture, because there are no distributional issues to speak of and (therefore?) no great ideological causes to worry about.

Instead of deliberative politics, the Culture has the logical extension of machines as citizens: because machines are orders of magnitude more efficient than humans, the Culture is "ruled" (using this word loosely, because the average Culture citizen doesn't encounter authoritative commands very often) by ultra-advanced machine intelligences called Minds. Each of the Culture's starships and orbital habitats has at least one Mind running the show, keeping all of the systems in operation with a fraction of its capacity while looking after all of the inhabitants of the ship or orbital as a combination nursemaid and interlocutor. Minds do this not because they are programmed to—no human has *programmed* a Mind for thousands of years, and Minds are grown by other Minds in ways that simply wouldn't make sense to the human mind—but because, as citizens of the Culture like other citizens, they find fulfillment in doing so. (Minds not involved in taking care of other beings so centrally spend their time directing the Culture's activities toward other civilizations, pondering the nature of reality, or, on those rare occasions when the Culture encounters a genuine rival, making sure that the Culture prevails.)

The Banks solution to the problem of machine actorhood, then, is to treat any machine sufficiently powerful to simulate intelligent thinking as an actor with as much status as any other and not to insist on maintaining the human/machine boundary. Inasmuch as quantum AI opens the door to just such computers and androids and other beings, this approach would suggest looking for signs of intelligence in quantum computational systems and then seeking to engage with and educate it. Even though the Minds have long since surpassed the humans who constructed their ancestors, Banks presents a world in which those Minds retain some traces of the society that gave birth to them: Culture Minds are not exactly like the artificial intelligences created by other galactic civilizations, in subtle and profound ways. One might say that *because* they have not been treated as servants or threats, they have not become so, and their "rule"—or perhaps better, their *arrangement of circumstances*—benefits the Culture as a whole and all of the beings within it, both biological and mechanical.

Quantum AI may therefore be a gateway to the surpassing of human beings by their creations, but this may not be a bad thing—and it may certainly be better than a war against the machines, a war that human beings could well not win. Better, perhaps, to work to ensure that our intelligent creations take forward the best we have to offer, and not the worst.

CHAPTER

# 4

# Data Privacy, Security, and Ethical Governance Under Quantum AI

Sarah Pearce, *Partner, Huntons Andrews Kurth*

Data privacy, security, and ethical governance issues have long been raised in respect of the use of artificial intelligence technology—and can to some degree be managed, provided certain precautions are taken. The convergence of quantum computing and AI takes the challenges to a whole new level; the speed with which these technologies are advancing means legal theories, policies, laws, and regulations that had evolved (or were in the midst of evolving) to harness the issues of AI that are already out-of-date. Let us consider the impact of the technology on the social and regulatory impacts it may have on society and a body of regulation already struggling to keep up with technology.

## Insurmountable Data Privacy and Cybersecurity Issues?

The data privacy implications of AI are vast, particularly if we consider the requirements of the multitude of legal regimes in existence around the globe and those that are on the horizon. In addition to national legislation, there is often an industry overlay of regulation that needs to be considered, and we are now seeing an independent set of supra-national rules and regulations specific to AI itself coming into play. Forty countries have adopted the OECD's five AI principles, for example, and in 2019, the G20 expressed its support of these principles. The supra-national attention given to the regulatory framework of this technology is testament to the potential it offers; yet this is all based, so far, on the "simple" or "classical" form of AI, notably in ignorance of AI's convergence with quantum computing.

Compliance in such an expansive and diverse world of regulation can seem overwhelming. There are, however, a few key themes emerging from lawmakers and enforcers: these include fairness, lawfulness, and transparency; data minimization (limiting data collection to only what is required to fulfill a specific purpose); rights of individuals; and security. We consider each of these next.

The complexity of even the most basic AI technology often means it is difficult for a company to explain clearly how it uses personal data—and indeed, the particular algorithm or technique deployed can cause the use to change over time, likely becoming increasingly complex. It will often be difficult to identify a single lawful basis for the processing taking place, one of the key requirements of the European regime. How can a company rely on consent, for example, if it is unable to explain clearly the processing being performed now and in the future as the algorithm evolves? Such potential repurposing of personal data in unexpected ways, using complex algorithms that enable conclusions to be drawn about individuals with unexpected and possibly unwanted effects, could pose a threat to individuals' personal data. The key is to be transparent about what data is being collected and used—and how. So, if this is difficult enough to achieve by the very nature of existing AI technology, a looming question remains as to the extent to which quantum computing could assist with meeting transparency requirements.

Most uses of AI currently rely on the collection and analysis of large volumes of data, with companies at risk being accused of taking their collection activities toward the excessive and being asked to explain whether and why they are retaining the data for longer than may be perceived necessary. While any potential "overuse" of personal data is most likely with a view to improving and further replicating algorithms, this is not something that is welcomed bearing in mind the data minimization principle at the core of data privacy regulation. The convergence of quantum computing with AI, arguably, brings increased efficiency, and one might hope that means the same result being achievable with a lower volume of personal data being collected and retained.

Bringing quantum computing together with artificial intelligence might seem to bring benefits in meeting—or at least dealing with—data privacy requirements and facilitating compliance therewith. However, the rights of individuals that exist over their personal data after the data has been absorbed and processed within AI applications remain a stumbling block. How can the information be retrieved, extracted, and provided to the individual, for example? Unfortunately, the limits of quantum AI become apparent here; it is difficult to see how this hurdle can be overcome or even facilitated by the convergence of the two technologies. Increased transparency may assist, but more critical is that companies ensure they are developing their technology with such rights (particularly rights of access) in mind from the very start, building in appropriate security mechanisms. As the emergence of quantum AI becomes a reality, it may be possible to foresee a state where the concept of "privacy by design and default" (i.e., building in such rights to algorithms from the get go) becomes a reality that can be implemented practically—and efficiently.

Of course, we cannot consider data privacy issues without considering security, and cybersecurity in particular, which takes on profound importance in the context of AI technology. Effective controls are required to handle the mass volumes of data used in traditional AI models, and it is with hope that we can look to the convergence with quantum computing and its increased efficiency that such controls can be strengthened. "AI poisoning" (manipulations of AI training data impacting the quality of algorithms and the decisions they produce) has sadly become a recurring feature in the more classical AI models, so any enhancement in the protection of the integrity of algorithmic development processes with appropriate security controls and procedures is to be applauded. The security of autonomous systems based on AI technology is of course critical to ensure safe operation in the physical world. It hardly needs saying, but poorly secured autonomous vehicles or drones, for example, can create significant risk to human safety and property and attendant liability. Malicious actors, ransomware attacks, and cyber incidents of all shapes and forms are on the rise. Scientists will be called upon to exploit the technical advantages brought about by quantum AI, with a view to ensuring protection from such attacks that have the capacity to compromise the safe operation of AI-powered technological advancements with such catastrophic effect.

## Ethics and Good Governance Structure: Do We Ensure Outcomes Free of Bias in "Opaque" Technology?

Fairness and consistency of decisions made by algorithms is not confined to data privacy compliance matters: as use of AI technology is reaching deeper into consumers' lives and spreading to more corners of industry, there is an increased expectation of fair and consistent application. Much criticism has already been voiced vis-à-vis the real-life issue of AI-driven services delivering outcomes that are inherently biased. The pressure is on for businesses implementing such technologies to ensure bias-free use, whether that is on the basis of gender, race, or other categories. On the one hand, it may appear, superficially at least, that decisions rendered by algorithms are even more opaque when quantum computing is interwoven with AI than when the more classical algorithmic decision-making is in operation. On the other hand, it may be that the injection of quantum computing into the classic AI model drives greater efficiencies with an increased chance that decisions may be rendered free of such bias.

The Centre for Data Ethics and Innovation (CDEI) recently published its review of the presence of bias in algorithmic decision-making, which looks primarily at how algorithms can be used to promote fairness rather than undermine it. The report acknowledges the very real risk of bias in the deployment of AI technology. Data used to draw general conclusions can result in excessively high error rates when based on a small dataset, and this can have significant adverse consequences when implemented in certain sensitive arenas. The report also recognizes the transparency issue but notes that any requirement as to transparency must be appropriate to the context and strive to meet the reason for its being: to promote understanding and trust. It would not be appropriate (or useful, nor would it resolve the issue in the context of algorithmic decision-making) to publish the computational algorithm; most would either not understand or deduce incorrect meanings. The report makes certain recommendations as to how to meet transparency requirements, which essentially amount to providing a description of the algorithmic decision-making that is accessible to the average person: an explanation that is easily understood and deals with possible concerns.

The critique cannot stop there, however. We need to go beyond the simple question of whether decisions are "fair" to whether the outcomes are actually "right." The CDEI review includes a helpful reminder of this point: "The issue is not simply whether an algorithm is biased, but whether the overall decision-making processes are biased. Looking at algorithms in isolation cannot fully address this."

This then leads to a consideration of quality-control issues. The tension between machine-driven decision-making and the legal system's insistence on accountability is real and looks set to rise. Getting the transparency bit right is of course key, but algorithmic accountability is critical for widespread deployment and success. One would hope that quantum AI will enable more built-in mechanisms to facilitate transparency and accountability, but it is hard to see how it can completely dispel the concern. We are seeing global efforts to regulate here: the Canadian authorities have, for example, developed an algorithmic impact assessment: a questionnaire designed to assist organizations with their analysis and manage risks associated with the development and deployment of AI technology.

To protect against civil liability, regulatory scrutiny, and reputational harm from unintended outcomes across a wide variety of AI usage, governance structures are required. Again, could quantum AI assist? This seems unlikely given the widespread consensus among specialists and commentators (not to mention the general public) that any good governance structure should be designed and overseen by humans—or at least incorporate a human component at some point in the decision-making process, even if limited to a particular level of algorithmic deployment. The CDEI Review provides useful insight here, and while it rightly acknowledges that the components of an effective governance strategy will differ by industry and application, certain elements will be common. In its view, leaders (humans) should remain accountable for understanding the capabilities—but also the limitations—of any AI technology, perform impact assessments as appropriate, and ensure appropriate levels of human involvement. Regardless, a good governance structure (subject to the industry/application differentiation mentioned earlier) comprising certain key criteria is vital.

Externally, we should expect to see statements from companies deploying AI technology that evidence commitments to quality outcomes, consumer privacy, safety, unbiased and fair decisions, and an appropriate level of transparency around the use of AI technology in question. Notices will of course be required in line with consumer protection and data privacy requirements (and will need to include a means for consumers to have some recourse if they believe there has been unfair treatment as a result of a decision made on the basis of an algorithm), and businesses should ensure these are given at the appropriate moment in time (usually prior to deployment).

Internally, organizations need to ensure they have in place robust quality-control mechanisms to assess the outcomes of any algorithmic decision-making, all with a view to ensuring the absence of bias. A risk-based approach is key here: use of AI technology in the health arena, for example, is likely to require greater control than that in the context of consumer marketing. A wealth of policies and procedures will also be required to ensure any development and deployment of such technology is delivered with the utmost impartiality. At a minimum, businesses using algorithmic decision-making in their operations, or their products/services, might want to consider internal requirements as to qualifications of individuals closely engaged with its operation.

The benefits of increased AI usage in society are, even if not without risk and the subject of much criticism, becoming more broadly accepted, with UN groups even believing it can help to promote the public good in facilitating the achievement of the UN Sustainable Development Goals by 2030.

However, that is not without challenges. Data privacy and security issues are inherent in the technology—and the resulting problems are real and serious. Robust and accountable privacy and security governance frameworks, solidly founded on a set of core ethical principles, will be vital to the success of any widespread rollout of the technology.

Taking the words of the insightful CDEI report, "What is clear is that, given the pace of change, and the wide range of potential impacts, governance in this space must be anticipatory." Ideally quantum AI enhances the technology's capabilities such that it is able to thrive amid an appropriate level of regulation to counteract potential unacceptable issues.

CHAPTER

5

# The Challenge of Quantum Noise

Philip Johnson, *Associate Professor in the Department of Physics, American University*

Quantum computing and artificial intelligence share an important characteristic—the ability to produce results not available by other methods and that are difficult to independently verify. Explainability and verification aren't always hard. For example, factoring numbers into their composite primes is a problem with both explainable algorithms and where we don't have to trust candidate solutions because we can easily check if they work. But complex learning algorithms trained and optimized over large datasets and used to solve problems of very high dimensionality are inherently hard to verify and explain. Yet for many problems, it is important to not only efficiently obtain a result but to understand why the result was produced and to verify its reliability. It is dangerous to simply trust complex solutions to complex problems.

The tendency to be black boxes arises separately for both artificial intelligence and quantum computers, but achieving explainability and verification will be significantly harder when they are combined as quantum AI systems for at least three reasons. First, the states of quantum AI systems live in exponentially large Hilbert spaces with fantastically high dimensionality such that, for even modest-sized quantum AI systems, we will simply never be able to directly sample more than an infinitesimal fraction of the system's state space. Second, quantum mechanics forbids direct interrogation of intermediate states of quantum AI systems lest we disrupt their computations, resulting in a form of unexplainability enforced by the laws of physics. Third, the properties of quantum noise inside quantum AI systems are much, much harder to mitigate than the effects of ordinary, classical noise. Nevertheless, as long as it is believed that quantum AI systems offer competitive advantages, the temptations, pressures, and incentives to use them, even in domains where we don't have methods to verify or reasons to trust their results, let alone understand their reasoning, will be strong. Here I focus on the risks from quantum AI systems mistaking quantum noise for meaningful patterns and making decisions that effectively amplify highly correlated noise to a macroscopic scale.

The power of quantum computers derives from creating and exploiting quantum entanglement between quantum bits (qubits). When there is noise in a quantum computer, gate operations or other interactions between qubits spread quantum-correlated errors throughout the system, unless there is fault-tolerant quantum error correction to prevent it. Although it is difficult to predict with confidence, it will likely be many years, possibly even decades, before we have true large-scale fault-tolerant, error-correcting quantum computers.[1]

Fault-tolerant quantum computers require extremely challenging engineering to make it possible to preserve and control quantum entanglement for a long enough time to perform complicated quantum algorithms. To give a sense of the magnitude of the challenge, one recent analysis assuming a 1 percent physical qubit error rate estimates it may take 1,000 to 10,000 physical qubits to perform collectively as a single effective fault-tolerant logical qubit. The number of physical qubits required per logical qubit could be higher or lower, depending on the details of the error types and rates, but unless we get lucky (e.g., successfully develop an advanced, scalable qubit, architecture, and control system that is highly immune to errors), it will take many physical qubits to implement even a single fault-tolerant logical qubit. The current state-of-the-art quantum computers have less than 70 physical qubits and errors rates on the order of 1 to 2 percent, and to date, not even a single fault tolerant logical qubit has been demonstrated. Although the number of physical qubits will steadily increase, for example, IBM is targeting a 1,000 qubit quantum computer by 2023, it looks much more challenging to simultaneously achieve the small error rates and precision control needed for true scalable logical qubits.

Until we have scalable fault-tolerant quantum computers, we will continue operating in the regime of what John Preskill at CalTech has coined noisy intermediate-scale quantum (NISQ) computers, systems that utilize noisy qubits and gate operations and where correlated errors spread over time from qubit to qubit. The first quantum AI systems will therefore probably be NISQ-AI systems, systems that utilize a combination of both AI methods together with noisy qubits and gates. The first stages of this development are well underway. Classical machine learning algorithms running on classical computers are already used to passively analyze the data generated by NISQ computers and for actively controlling noisy quantum processors. Eventually, however, quantum learning algorithms will directly incorporate qubits into their hardware, and it is this stage of development that I am calling a NISQ-AI. An example of an NISQ-AI would be a neural network where some or all nodes are noisy qubits.

The power and usefulness of NISQ computers, let alone NISQ-AIs, is an open question. The controversies around achieving quantum supremacy, loosely defined as performing any calculation significantly faster than any realistic classical computer can, illustrate how difficult it is to quantify the capabilities of an NISQ computer, even when they are used to solve problems specially selected with the sole purpose of helping us characterize an NISQ computer's performance. For ever more complex NISQ-AI systems, it is easy to imagine this challenge only increasing.

The problem of noise in NISQ-AIs is that errors in the state of a single qubit spread throughout the system (in the absence of fault-tolerant error correction) leading to entangled, nonlocally correlated errors of increasingly high dimension. Distinguishing between noise and meaningful correlations is already challenging when analyzing classical data of sufficiently high dimension. Standard methods of analysis assume the statistical independence of noise sources such that averaging smooths away the random fluctuations that can look like false signals, leaving behind the true underlying signals. Classical machine learning algorithms, although they use nonlinearity to amplify signals and patterns in data, still rely on averaging to suppress the effects of noise.

These methods can fail for systems with nonlocal quantum noise because the random fluctuations involving even very widely separated qubits will be "spookily" correlated, breaking the usual assumption of spatially independent fluctuations. The challenge is that spatially correlated noise can look to AIs like meaningful correlations in data, and learning algorithms seeing those correlations will amplify them into convincing signals—ghosts emerging from quantum noise. Using standard diagnostics, they may be indistinguishable from meaningful patterns. Considering how often, even for purely classical data, noise is mistaken for meaningful information (the efforts to model and predict financial markets, political polls, and health data are familiar examples), it is not hard to imagine how much easier it will be to mistake amplified quantum noise, because of its unusual and highly nonlocal features, as meaningful signals.

As NISQ-AIs grow in number of qubits and coherence time, the task of adequately estimating the system's quantum state in detail will become intractable, which could in turn make determining whether nonlocal noise is being amplified as false signal unknowable. This possibility, given the inherent intractability of verification and explainability for these systems, could make using NISQ-AI systems for decision-making on important problems very risky. We find it challenging enough to avoid mistaking classical noise for true signals. However, quantum noise is more "dangerous" than classical noise precisely because it doesn't look like ordinary noise, with the danger compounded by quantum noise being resistant to classical risk mitigation strategies that generally assume that random noise is local and uncorrelated and that nonlocal correlations are significant.

Given these issues, why might we misuse NISQ-AI systems outside carefully constrained domains? Because after years of investment, together with the influences of personal and institutional ambitions, there will be strong incentives to use the technology for real-world applications. Because evaluating the performance, reliability, and risks of NISQ-AI systems is so difficult, the problem may evade a definitive community consensus, creating opportunities for individuals, organization, and nations to convince themselves that their system can provide a competitive advantage, while discounting potential risks. Because, as previously noted, false signals generated by quantum noise in NISQ-AI systems won't look like (ordinary) noise. And because mistaking noise for signal, and the closely related problem of mistaking correlations for causation, is as old as humanity.

The consequences from misusing NISQ-AI systems will obviously depend on how they are used. If the systems are used for exploratory investigations, research purposes, or well-constrained problems like quantum chemistry or material physics simulations, the consequences in the case of unreliable performance will probably be limited to, at worst, lost time. For example, incorrect simulation results could guide research in unfruitful directions. This risk is balanced by the potential benefits if these machines produce important results or help us toward developing more powerful future technologies. The dangers if NISQ-AI systems are used for critical decision-making, however, are more speculative but warrant examination. It is not hard to imagine, for example, NISQ-AI systems being used for high-speed trading in financial markets.

Consider, for instance, a system giving correct predictions often enough to incentivize and reward its continued use, but some unknown fraction of the time results in trades based on quantum noise. If this occurs at sufficient volume, could it inject correlated noise into markets at a scale that feeds market volatility in unusual and destabilizing ways? The 2008 market collapse is partially attributed to the failure of risk models that incorrectly assumed that correlated variables, such as between housing markets in separate geographic regions and between different types of financial instruments, were uncorrelated. The collapse provides a powerful lesson on the dangers in systems with strong, hidden correlations: the potential for unexpectedly large, correlated effects, and the inadequacy of risk mitigation strategies that assume uncorrelated variables. How much greater might be the risk when there are hidden correlations injected into markets that are truly undetectable?

Governments and militaries tend to be more risk averse than corporations and therefore may be less likely to use technologies like NISQ-AI systems in critical processes until they are adequately tested and verified. On the other hand, if NISQ-AI systems end up providing demonstrated advantages under what seems like adequate testing conditions, and if quantum noise rates are small enough that the dangers from correlated errors aren't often experienced, there will be the temptation to use these systems and discount the risk. NISQ-AI systems that mistake correlated errors as signal are effectively amplifying nonlocal quantum noise to macroscopic levels. Of course, noise and randomness are familiar parts of life, but nonlocal noise isn't. In the presence of significant nonlocal noise introduced by NISQ-AI systems, if they are operating in unconstrained ways, we will need to recalibrate how we think about and protect against chance. We could be creating a world where black swan events are far more common.

Before NISQ-AI systems are used for critical decision-making applications, policy makers will need to carefully examine the risks and benefits from using unverifiable and unexplainable systems and reconsider how our familiar intuitions about risk can fail in a quantum world.

CHAPTER

# 6

# A New Kind of Knowledge Discovery

Ramin Ayanzadeh, *Post-doctoral Fellow, Georgia Institute of Technology;* and Milton Halem, *Professor, University of Maryland Baltimore County*

At the same time as the first solid-state device (the transistor) was being developed at Bell Labs in the mid-20th century to replace vacuum tubes,[1] artificial intelligence (AI) was being conceptualized by a generation of scientists, mathematicians, and philosophers. In 1950, Alan Turing suggested two criteria for machine intelligence: memory for enabling machines to store and retrieve data, and reasoning (i.e., having the capacity to process data).[2] Since then, trends in doubling the transistor count, characterized by Moore's law, have catalyzed AI advancements. Nowadays, AI applications have access to not only large-scale memories but also high-performance computing (HPC) resources.

After decades of predominance, the era of Moore's law seems to be drawing to a close. Are we prepared for the end of this era? Can digital systems keep pace with ever-increasing demand for data storage and information processing capacity? The microelectronics industry (soon to be known as the nanoelectronics industry) is trying to identify new materials and devices to replace the 50-year-old transistor technology—including, but not limited to, nonclassical complementary metal-oxide-semiconductor (CMOS, such as new channel materials) and alternatives to CMOS (e.g., spintronics, single-electron devices, and molecular computing).[3] Although the microelectronics industry will continue to reduce the cost of electronic devices, there are theoretical and physical boundaries that limit classical processing devices' computing power. However, it is worth noting that AI applications can still benefit from the ongoing increase in classical memory systems' speed and storage capacity.

Transitioning from vector-based computation, in central processing units (CPUs), to matrix-based computation resulted in the emergence of graphical processing units (GPUs) that have reshaped the landscape for accelerated computing, namely, in realms of scientific computing, high-performance computing, machine learning, and Big Data analytics. In the same manner, custom design application–specific integrated circuits (ASICs) for extending computations (from vectors and matrices) to tensors (i.e., complex and higher-order objects) can provide a further disruptive capability. As an example, the tensor processing unit (TPU), by Google, is an accelerator for near-real-time deep learning applications with low latency that has demonstrated throughput improvements of more than 15 to 30 percent and power efficiency improvements of 30 to 70 percent over current CPUs and Kepler generation of GPUs, albeit lower precision computations.[4]

The supercomputing community tries to address the limits in shrinking transistors' size through parallelism (i.e., distributing data and processes), which has emerged as a new kind of race (or war) between the United States and China. Besides national security and economic concerns in developing the next generation of supercomputers, the near-term future of AI can mainly depend on the result of the supercomputing race. Hence, we can expect that the winner of the supercomputing game will achieve AI supremacy.

## The Post-Moore Era: Emerging New Technologies

Notwithstanding the advance of the computing power of the next generation of accelerators, there is a growing consensus that we will eventually need different types of computing machinery. In the post-Moore era, therefore, we will explore non–von Neumann architectures (such as zero-instruction set and single-instruction set computers) and nondigital systems (namely, quantum computers) for the emerging next-generation of accelerators. For example, neural processing units (NPUs) or neuromorphic chips—namely, the Neurosynaptic System by IBM, the SpiNNaker System by the University of Manchester, Intel's Loihi chip, and memristors-based systems—are zero-instruction set computers inspired by the human brain that have demonstrated a dramatic speedup in implementing AI models, more specifically deep neural networks.

In 1982, Richard Feynman proposed the use of quantum systems for simulating quantum processes.[5] But why does one need a quantum computer to simulate quantum processes when our classical silicon-based computers can solve all types of physics problems from relativistic to Newtonian models. The answer lies in the following three seemingly unreal properties that atoms exhibit when cooled down to near absolute zero.

First is the property of *superposition*. While a classical bit can be either 0 or 1 only, a quantum bit (qubit) is a two-level quantum system that can be 0 and 1 simultaneously (called *superposition*) with corresponding probabilities. Let's say one were to encode information into a quantum state or analogously into a quantum bit, whether the state is created by magnetic fields about a gallium arsenide chip or by trapping ions with lasers or forming arrays of cesium atoms at one-millionth of a degree, these states represented by a 1 or 0 would form the bit structure of a quantum computer. A quantum register with $n$ qubits can simultaneously be in $2^n$ arrangements. This means a quantum system with 100 qubits can have $2^{100}$ arrangements, a truly massive number that would enable unbelievable searches.

The second property known as *entanglement* is even stranger. Assume two such sets of quantum bits were correlated in Washington DC, and then one set of quantum bits is separated by the Pacific Ocean in a site in China. Then any operation on the quantum bits in Washington DC would simultaneously enact at the entangled quantum bits in China.

Finally, the third property is known as *interference*. Unlike classical wave functions that can interfere only with each other, in quantum mechanics, an individual particle can cross its own trajectory and interfere with itself. Quantum interference enables us to bias the measurement of a quantum register toward the desired outcome.

These three properties enable quantum operations to perform functions that are effectively impossible with classical computers. The challenge to implement such a quantum computer is that these states are so responsive to minor temperature, radiation, and vibrational effects that noise limits the time to conduct operations on the quantum bits to fractions of milliseconds.

However, we are confident that quantum science has the potential to emerge as a transforming technology, and quantum technology promises new capabilities in demonstrating advantage in several domains, ranging from problem-solving to sensing, communication, and simulation of quantum physical systems (such as in high-energy physics).

## NISQ Era: New Discoveries

Fault-tolerance quantum computing relies on continuous error correction; nevertheless, existing and near-term quantum computers cannot fully accommodate quantum error-correction techniques. Hence, until we can bypass several technological barriers, we are limited to explore noisy intermediate-scale quantum (NISQ) computers for exploring the quantum advantage.[6] It has been theorized that we will need thousands (or even millions) of physical qubits to achieve the quantum advantage, so we should not expect to be able to run any quantum algorithm on a near-term quantum computer. While this attitude sounds limiting, we should highlight that quantum computers are not exclusive for only running quantum algorithms for addressing certain types of (classically intractable) problems.

From an application point of view, we can be optimistic about exploring quantum advantage through applying NISQ computers on (at least) two broad classes of problems: optimization and simulation. In the circuit model of quantum computing, variational quantum algorithms (aka classical–quantum hybrid schemes) such as quantum approximate optimization algorithm (QAOA) and variational quantum eigensolver (VQE) can address optimization applications that are intractable in the realm of classical computing.[7] In fact, we can expect the next generation of NISQ processors (namely, cold atoms) to be capable of executing large and deep enough quantum circuits, which is the bottleneck for demonstrating the quantum advantage using QAOA and VQE.[8]

Physicists are also excited about using NISQ devices for exploring the physics of many entangled particles. We know that quantum computers can simulate any natural process, and NISQ devices provide a valuable platform for discovering new aspects of physical processes.[9] To this end, we can expect groundbreaking new discoveries when we have access to NISQ computers with a few hundred qubits.

While most current quantum artificial intelligence studies propose applying quantum accelerators to hard AI problems, NISQ devices are less likely to provide a disruptive capability in this area. Ironically, in the NISQ era we can expect AI to improve the fidelity of near-term quantum computers. Recent studies have suggested that machine learning/AI models can mitigate the measurement error, which is the most error-prone operation in most current quantum circuits. Since NISQ devices are susceptible to various error sources, a large number of trials are needed, and the correct answer is inferred based on the distribution of outcomes. In this context, deep classical networks can be trained to learn and mitigate the system noise of NISQ devices.

## Post-NISQ Era: Quantum Advantage

We are currently in the NISQ era where we are limited to small and noisy quantum processors, but we are confident that fault-tolerance quantum computing is on the horizon. Quantum artificial intelligence (QAI) and quantum machine learning (QML) are emerging fields that aim to leverage quantum computing for addressing certain types of problems that are intractable in the realm of classical computing. Although we do not expect NISQ accelerators to be the game changer in the realm of artificial intelligence, we do expect fault-tolerant quantum computers to be a turning point.

While most studies are optimistic about leveraging quantum computers to accelerate AI and machine learning applications, we should not expect quantum machine learning to fully outperform classical machine learning in the near term.[10] There are theoretical boundaries that limit the potential advantage of quantum machine learning over classical machine learning in some cases, while on the other hand, quantum machine learning can outperform classical models in other situations. Thus, the performance of quantum AI and quantum machine learning will mostly depend on applications rather than the model, so we need to find problems where quantum AI can be of practical use. Weather prediction is an example where we can expect quantum machine learning to be a transforming technology, and we also expect quantum-assisted predictive models to play a crucial role in finance applications.

In the biopharmaceutical industry, developing a drug product from an initial scientific hypothesis can take more than 10 years and billions of dollars before being commercialized. The vast majority of the time, effort, and cost are spent on experimental design and characterizations. Despite considerable progress in classical computers for modeling macromolecules such as proteins, ligands, and peptides, many molecular biology and biophysics challenges remain computationally infeasible. Numerically calculating the full electronic wavelength of a drug product is expected to take longer than the age of the universe, even on the current most powerful supercomputers. Quantum computing promises to speed up drug discoveries exponentially, and we are optimistic that machine learning can expedite the achievement of quantum advantage. It is worth highlighting that similar concepts can be applied in material science to predict characteristics of new unknown materials.

The era of quantum knowledge discovery will open up a whole new world. For example, the biopharmaceutical industry will no longer be considered as a (very) high-risk sector, and we can expect novel discoveries to flourish. Besides its economic impacts, we can expect a notable improvement in life expectancy through new drug discoveries and novel early diagnosis systems. Quantum science and technology can also enable us to predict and prepare for upcoming climate changes, and also address complex global problems such as pandemics.

However, it should also be noted that quantum technology will likely increase the gap between developed and developing countries from the knowledge discovery perspective. In the era of quantum knowledge discovery, access to premium quantum resources (not just the hardware, but also the algorithms and scientists to run them) will play a crucial role in scientific research studies. Equitable access to quantum technology is important on many levels, and without it we run the risk of further entrenching the global inequalities in existence when the quantum race is won.

PART

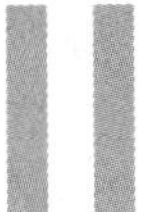

# Economic Impacts

# CHAPTER 7

# Quantum Tuesday: How the U.S. Economy Will Fall, and How to Stop It

Alexander W. Butler, *Associate Director, Quantum Alliance Initiative, Hudson Institute*

"Even the most utopian of today's visionaries will have to concede that the mere existence of modern technology involves a risk to civilization that would have been unthinkable twenty-five years ago."

—Herman Kahn, *Thinking About the Unthinkable*, 1962

New York, New York

Friday, November 3, 2028*

---

*This is a work of fiction. These events are set in the future, and any resemblance to actual persons, organizations, or events is coincidental.

Back at the start of the new millennium Wall Street raved about the dawn of artificial intelligence. Not only was it meant to improve everyday life, but these "smart" algorithms were set to revolutionize everything from high-speed trading to the Federal Reserve's interest rate decisions. With the combination of the new generation of AI, paired with the dual powers of quantum annealers and classical supercomputers, a new industrial revolution had already begun. Medical research was now supercharged, and the traffic grid was now operating smoothly enough to enable a full rollout of completely autonomous vehicles. The financial sector was likewise undergoing a revolution. At first, the technologies were applied to front-end operations and customer service. But now automated smart trading was in full swing. Traders were becoming merely a safeguard against machine error, not that the over-engineered software needed much oversight. The only real role any of us traders now played was in curating and calibrating our models and algorithms. So, my day-to-day job had gone from actually analyzing the markets to analyzing the code for my firm's new AI-enabled trading program. Like the rest of the Street, my team and I were riding the booming bull market. And with the emergence of the new universal quantum computers, the future never looked so bright, at least for the select banks and firms that could afford them.

Yet, a key concern raised about all of this was the equality of it all. Equality—how it had become the word of the decade. Now even quantum was coming under the auspices of the self-declared guardians of equality in the Senate. The Financial Oversight Committee had a new focus: quantum equality—or the growing inequality between those countries, and even companies, that could afford the powerful machines and those that were beginning to get left behind in the new, quantum industrial revolution. And so, the focus shifted from safeguarding the development of quantum computers to regulating those who already had them.

JPMorgan was the only major player in the United States that could afford to privately fund the development of its own universal quantum computer—and it showed. Its returns had skyrocketed to nearly double the level they had been before the announcement of the new fault-tolerant system. Everyone else on the Street was using some version or another of D-Wave's quantum annealer machine, which did the job well enough in conjunction with classical supercomputers for their modeling and Monte Carlo simulations. All the while, smaller banks and trading houses were stuck renting time on the big firms' machines. This proved to be a big concern for the regulators in Washington. Not only did the big players—the big five banks, the massive mutual funds, and others—have an advantage in terms of performance but also publicity. Everyone, even the retirees, wanted to bank and to invest where the huge new quantum computers were. Such was the promotional power of the firms.

The public—hell, even I—barely understood the power, nor the physics, that made these quantum machines tick. Most of my quantitative finance courses had covered coding for the algorithms, but none of us analysts knew much beyond the superficial about quantum computing. But while everyone on the trading floor understood that, between the deep-learning algorithms and the quantum annealers, we could optimize and automate our trades down to fractions of milliseconds, the public understood what the big bank commercials and the articles in the *Financial Times* and *Wall Street Journal* told them to be true—the "quantum advantage" was real.

The advantage played well—so well, in fact, that the main Wall Street hawks on the Senate Financial Oversight Committee drove the point home on all the Sunday talk shows. Some on the fringes even called for regulation on the machines themselves. Most, however, focused on the usage. But that wasn't the problem as all the players on the Street understood it. All the trading houses were using some variant of the same machines, all of which ran some iteration of the same deep-learning algorithms. This was especially true of the smaller firms that were relegated to "time-sharing" their modeling capabilities.

Unfortunately, this never got explained to the Senate staffers, nor the general public. The "quantum [in]equality" headline had gotten enough attention that all the big names—CEOs and CITOs from the Street to Silicon Valley—were called down to the Hill for testimony. They even called the Fed chair and the New York Fed governor in for remarks.

The Federal Reserve was one of the biggest and earliest adopters of AI, first, and then quantum computing. Like everywhere else, the combination of these technologies had really amped up its modeling capabilities, with the ability to run some of the most complex econometric and macroeconomic models—some with millions of variables—in a matter of seconds. Suddenly, with this newfound computing power, the Fed was actually able to accurately predict macro trends instead of just responding to them. Apparently, it was even using some complex optimization algorithms to aid its interest rate decisions. That the governors at the Fed were proponents of the "smart quantum advantage" was an understatement. The New York Fed governor had even gone so far as to be quoted saying "The convergence of AI and quantum technologies would usher our economy toward an era of heightened efficiency and equitable growth"—primarily for the financial sector, he didn't add. His comment attracted a lot of mostly negative attention, which was the point—the Chair was trying to shift the focus away from the Fed's unveiling of the new, U.S. government–backed cryptocurrency.

The media was calling it the "quantum hearings," and it just so happened that day two of the hearings fell on Tuesday, October 31st, 2028—a week before the U.S. Presidential election. All of the biggest names and brightest stars in the FinTech world—exactly the people who might have been able to stop what was about to unfold—were gathered in a Senate committee room, cut off from the outside world. It was Halloween, and sure enough the Senators, Fed officials, CEOs, and CITOs would get one hell of a scare—we all would.

But with all the hype and scrutiny surrounding the technologies, our government officials overlooked the malevolent potential the convergence of these two technologies created. We thought we had built a strong and resilient "smart" economic and financial system, one that would not only prevent another financial crash but even survive another pandemic lockdown like the one back in the early COVID days. But like the general who prepares to fight the previous war, the Fed was fighting to prevent the previous stock crash, the previous recession. Despite all the safeguards, against both human and machine error, somehow, somewhere, some junior staffer on the Senate committee forgot to mention that even the most secure system still relies on a simple and ancient device: a key.

While all those testifying were focused on these benevolent—and beneficial—applications of quantum AI, and while our elected officials focused on regulating equitable access to that same technology, we as a country seemingly forgot two important aspects of our new high-tech reality. First, the quantum capabilities that powered the applications, even the new deep-learning applications themselves, enabled even just a quantum annealer to run some very computationally difficult optimization problems. One specific algorithm, Shor's algorithm, had been reformatted as an optimization problem with the help of machine learning. Originally developed in the early 1990s as a database search algorithm, Shor's had been consistently upgraded and refined. But it just so happens that this algorithm is particularly adept at cracking certain encryption regimes—including those like RSA-2048, among others, upon which all of our cyber reality still relied.

While this issue had apparently gotten a lot of attention eight years ago—well before I began programming high-speed trading algorithms—we as a society were told not to worry. Evidently our techies in government had finally standardized and approved some new quantum-safe encryption that they had begun rolling out earlier this year. In theory, the rollout was early enough, but the second aspect we neglected was the simple fact that we weren't the only country to have access to these quantum-AI technologies.

So, we went about building a financial system that didn't just rely on quantum AI, but one that virtually ran on it. It was our operating system, and as the Senate hearings resumed after the lunch recess, we became all too familiar with our fatal "zero-day" flaw. The foundation of the technologies' benevolence rested on its applications and access to them. We would soon realize that we no longer controlled either. And so, it was overlooked at the hearings, the true danger of it all—Shor's algorithm, and our own eagerness to employ a technology without safeguarding our own systems against it.

I had just gotten back to my Bloomberg terminal after a quick lunch when the first ripple silently began to break. While everyone on the exchange floor had the hearings up on one of their many monitors, most traders were glued to their desktop readouts from their automated trading programs, all of which ran predictably on some version of the same AI, trained by some variant of the same historical datasets. Today was a big day for my team, however. I had spent all weekend "training" our flagship AI trading algorithm to a new dataset. This "Big Data"—an extremely vast and largely unorganized dataset—was the mother of all datasets. A true gold mine, our firm had just acquired it from a Chinese tech giant that specialized in social media apps. Costing us undisclosed millions, it promised to bring in tenfold in trading profits. With the new data, our high-speed program would "intelligently" watch the market and make real-time trading decisions based on historical precedent, and now, unique to this dataset, it would integrate revealed preferences of consumers from the Chinese firm's hit social media platform. Our program did this thousands of times a second, actively processing, learning from, and reacting to even the most minute changes in the global economy—in such a high volume of trades to not only outsmart and beat other investment algorithms to the punch, but in effect actually build and direct the market's momentum.

In essence, the markets are not truly a reflection of the economy, but rather people's expectations thereof. Consequently, the markets react to their own expectations, creating a sort of feedback loop—an effect exacerbated only by our AI-driven algorithms. As a result, the entire financial system ran off momentum and perceived momentum. Generally, these trends were only loosely interconnected. But on occasion, given the right trigger, different momentums could meet in a dangerous fashion. A single ripple in the market could meet just the right momentum to create a confluence of amplitudes—a tidal wave in the system. And as I sat back and watched my algorithm run, little did I know, a stone had been tossed into the financial pond.

It was 12:45 p.m. on Halloween when that first ripple appeared. The markets had been quite fluctuant, but that was to be expected this close to a Presidential election. To my satisfaction the data training had worked out, and the program played the election jitters well. We were up big on the day. That was until the Industrial & Commercial Bank of China suddenly, and seemingly irrationally, liquidated its entire U.S. Treasury holdings. A move against the dollar at such a magnitude was historically unprecedented, and consequently the majority of trading algorithms stuttered—their machine learning hadn't been taught how to respond to the Chinese bank's sell-off. But all of the other trading houses' algorithms were trained on American-curated datasets, whereas my firm's program ran on the new Chinese dataset. Shockingly, the program responded and did so quickly. The ripple had gained speed, and unbeknownst to me, it was about to gain force thanks to my data training.

Before I could react, or even process a productive move, the algorithm had sold off all our T-bills and dumped vast tranches of blue chips, which were evidently exposed to a weak dollar, favoring heavy metals and metal futures. At first this had seemed like a perfectly rational move but still one that should have required my human approval. Something or someone had automatically approved the trade, and it hadn't been me. I instantly stood up to flag down my supervisor, but before I had gotten their attention, the ripple had met the momentum it needed. At 12:47 the trading floor erupted into chaos as the Twitter feed on one of the big screens read out an AP report: "Run on the dollar underway in global financial markets." And so it began.

In physics it's called *constructive interference* when two waves pass through each other and their amplitudes converge into a single, larger wave. The ripple generated by the dumping of Treasury notes, amplified by my newly trained algorithm, converged with the force of the Associated Press' tweet. Before any of the human traders could process the moves, or even brace for impact, all of their automated trading programs made the same predictable moves—selling off anything and everything. At first it was to hedge against the weakened dollar, but by the time AP's Twitter account tweeted out announcing they had been hacked, a general sell off had begun.

By 12:52 p.m., as my team scrambled to figure out how the program had bypassed the approval protocol, the bell on the NYSE finally rang. The "speed bump" had automatically triggered the Exchange's circuit breaker to halt trading after the S&P 500 had dropped 7 percent—a collective sigh, but not for me. My supervisor had joined my efforts to scramble to fix the program. We soon realized that the trading program hadn't altered itself to bypass the needed human approval for our first moves—it had in fact been approved, and apparently legitimately so.

By this point word had already spread. In Washington DC, a Senate page had emerged in the hearings, and the Fed chair was permitted an early exit to handle the developing chaos on Wall Street. The "speed bump" wouldn't last long, after all, and the Fed needed to step in to shift the momentum and try to save the dollar. Per the suggestion of its quantum computer's model, the Fed issued an emergency authorization for a full one-point hike in the discount rate. Surely, they thought, this would quell the storm. None of us realized just how large the wave had become.

Surging past the "speed bump" and over the storm walls put up by the Fed, the wave of sell orders continued to flood the exchange. Only adding to the chaos was an announcement, just after 1 p.m., by the NYSE board that the exchange itself had been hacked. Someone had altered the ticker price program that fed information to the floor and news agencies alike. Now, as the AI-enabled algorithms continued to sell off their assets, the information they were processing and reacting to was unreliable. None of the algorithms knew how to appropriately respond. Our "smart" algorithms had gone haywire, simply adding to the momentum wreaking havoc on the U.S. financial system.

Like the Hindenburg, engineering failures built into the system sealed the fate of the once great American economy. At first it seemed like a fluke in the high-frequency trading algorithms. Little did we know that was just the spark. The ripple that had begun with a run on the dollar, augmented by my hacked algorithm and a hacked tweet, had evolved into a tidal wave—and it was about to crash over the entire financial system. As markets were forced to an early close at 1:34 p.m. on that fateful Tuesday, the selloff had transformed into a general run after one of the largest investment banks had become insolvent. The general selloff on the NYSE had been so quick that it failed to even trigger the level 2 circuit breakers. Erasing billions of market value in a matter of minutes, the automated trading algorithms, now joined by frenzied human traders, flew by the second "speed bump" before the level 3 breaker finally halted trading. Although initiated by the Dow's loss of 20 percent of its market value, the entire stock market went ablaze in a spontaneous inferno—many other individual stocks had lost twice as much value. Now, with the markets closed, the panic began to spread like wildfire. The general public was scrambling to salvage the remnants of their savings accounts, many of whom had already lost the majority of their investment portfolios to the failing markets.

Yet, just as it seemed that the wave couldn't get any bigger, PNC Bank announced that it was unable to remit payments due to another "technical glitch." Fearing further flashback, they failed to admit that their access to outgoing Fedwire payments had been administratively cut off. That PNC was scheduled to transition to the new NIST Quantum Safe standard in a week was now irrelevant. Soon the liquidity crisis began to cascade throughout the entire banking system. Even those that had transitioned to the NIST standards were compromised by the financial contagion. The entire banking system was an interconnected chain that held up the American economy. Now it became clear that PNC was the weakest link, and like a daisy chain, the unraveling of that one link inevitably led to the unraveling of the entire chain. The entire U.S. banking system was now failing.

By market open on Wednesday, the entirety of the U.S. economic system had gone up in a virtual blaze. That the dollar had lost half of its exchange value in a matter of minutes was now an afterthought. Now, after the news had spread across the foreign markets, overnight trading had reinforced the dollar's demise. Central Banks in Europe and Asia had all raced to dump their U.S. dollar holdings to preserve the value of their assets. The People's Bank of China was more than willing to extend its money supply, and now the Chinese RMB was quickly becoming the new global reserve currency. While the momentum in the markets is driven by expectations, the general macro-economy is, in part, driven by the momentum from—and the confidence in—the markets. The expectations that drove the U.S. markets had turned into nightmares. Now, the entire American economy had all but failed. And with ATMs and online transactions now impossible, public transactions were frozen as the national economy came to a standstill. The wave had finally crashed.

It would take government engineers a further 12 hours to realize that the RSA-secured systems had been broken into from a chain of virtual private networks. The engineers performing the post-mortem became more akin to forensic accountants chasing tax fraud through a network of shell corporations. Eventually they found recurring traces of the same Chinese IP address used to gain access into both my trading algorithm and PNC's servers from the safety of Hefei, China—home to the Chinese National Laboratory for Quantum Information Sciences and now the birthplace of the ripple. While we had built quantum supercomputers for medicine and finance, they had built theirs as a master key running Shor's algorithm, and now after the financial tidal wave had crashed, it was becoming increasingly evident that the Chinese Communist Party had employed their quantum-intelligent machines to break the back of the American economy. And thanks to the Chinese quantum computers, it all appeared authorized.

Bloomberg originally described it as a "technical anomaly," for the system would surely adjust and the markets correct, we were all reassured. Yet as the markets opened Wednesday, the whole trading desk watched in disbelief. The *Journal* was the first to coin the name, but by 10 a.m., as the Federal Reserve Bank, which had overextended itself trying to keep the banks afloat, failed and the President declared a state of emergency, the whole world would come to refer to the "anomaly" as "Quantum Tuesday." The flames that left the national economy in ashes soon spread to storefronts. The flood that first hit the trading desks soon ran into the streets. By Friday, the flood of riots had ebbed, giving way only to the flow of bread lines that stretched entire city blocks. We thought we had done everything right. But we forgot we were not the only country with this technology—and they had used it as a master key, a virtual locksmith key. The entire financial system relied on the perception of security, the naïve belief that if you locked away even a single dollar in the bank that it was safe. That we forgot to lock the vault no longer mattered, Beijing had already gotten away with the largest bank robbery the world had ever seen.

## Analysis

The scenario just described is hypothetical, yet one that may plausibly come to fruition all too quickly. Artificial intelligence and quantum computing represent the next frontier of information processing technology. Together, they hold the potential to answer some of the most daunting problems facing humanity. Yet, the particular properties that the amalgamation of these technologies will bring to benefit society at large also hold the potential to unravel the "unsolvable" mathematical problems that underpin today's public encryption systems. Beyond the capability of even the most powerful classical supercomputers, these complex mathematical problems protect vital data and networks, from banks and financial markets to air traffic control systems and the power grid—not to mention our government's most protected information and classified secrets. The dual development of artificial intelligence and quantum computing, and the resulting emergence of quantum AI systems, will enable our adversaries to decrypt and deconstruct even our most secure infrastructure systems, including our financial system underlying the national economy. We are already seeing many infrastructure attacks using existing technology, and it would be naïve to think they will not become even more common—and potentially much more devastating—following the development of quantum AI.

This is because current encryption regimes rely on the computational difficulty associated with the factorization of immensely huge numbers, a problem that classical computers cannot solve in a practical amount of time, if at all. But this kind of factorization is a skill in which future quantum computers will excel. Utilizing the power of long-existing quantum algorithms such as Grover's or Shor's algorithm, quantum AI technology will be able to break even the most advanced digital "locks" that we currently know about. This capability leaves most public key encryption—such as AES-256 or even Elliptic Curve Digital Signature Algorithm (ECSDA), used to secure cryptocurrencies and blockchain—exposed to quantum decryption. Especially exposed is a particular cryptographic system upon which large swathes of our digital world relies, RSA-2048.

Whereas it would take even the most powerful classical supercomputer some 300 trillion years to crack RSA-2048, this computational security evaporates in the face of a large-scale quantum computer. By reducing the computational difficulty of integer factorization from exponential complexity to only polynomial complexity, Shor's algorithm can theoretically crack 2,048-bit encryption in only 10 seconds. Nonetheless, this estimated time-to-break is still theoretical. Such a rapid computation would require a universal quantum computer with 4,099 perfectly stable qubits—an engineering feat that is years, if not decades, away. Yet, there are a number of scientific advances and technological innovations that continue to close the quantum-RSA gap. One such development has drastically reduced the number of qubits needed to effectively run Shor's algorithm.

While as recently as 2012 it was believed that a quantum computer would need one billion stable qubits to crack RSA, that number was reduced to only 20 million *noisy* qubits in May 2019.[1] However, further advancements in quantum annealing and reinforcement quantum annealing, the latter a sort of quantum artificial intelligence that is trained on an annealing device, hold the potential to decrypt 2,048-bit integers far sooner than a universal quantum computer. While not a universal, large-scale quantum computer itself, quantum annealing devices are highly capable at solving special optimization problems. Crucially, it has been shown that the factorization of complex integers can be re-formulated to run on quantum annealers as such an optimization problem.[2] Further exacerbating this trend is the fact that advances in AI are likely to advance developments in quantum technologies, and vice versa. This all culminates at a dangerous convergence, one that is likely to be met sooner rather than later.

Thus, although a majority of the literature on the quantum threat focuses nearly exclusively on the dangers posed by universal, large-scale quantum computers, the most imminent threat to society's encryption problem lies at the intersection of quantum-enabled AI, quantum annealing devices, and existing supercomputers. The gravity of this annealing threat was revealed in January 2019 by a group of Chinese quantum scientists who demonstrated that integer factorization problems could be reformulated as quadratic unconstrained binary optimization (QUBO) models—precisely the sort of optimization problems that quantum annealing devices are well suited for. This QUBO formulation enabled scientists to utilize an AI-improved quantum algorithm to factorize larger integers with fewer annealing qubits. Employing D-Wave's hybrid quantum/classical simulator *qbsolv*, the team successfully factored 1,005,973—a 20-bit integer—using only 89 error-tolerant qubits. This quantum integer factorization record has since been surpassed by another Chinese team in April 2020.[3] These breakthroughs confirm that "quantum annealing machines, such as those by D-Wave, may be close to cracking practical RSA codes, while universal quantum-circuit-based computers may be many years away from attacking RSA."[4] If this current rate of development of quantum AI is to continue, society will be facing the quantum decryption threat within the decade. While a universal quantum machine capable of running Shor's algorithm is some 10 to 15 years away, the annealing threat is much closer—and of equally devastating potential.

Yet, there are currently efforts to mitigate the severity and meet this imminent threat. U.S. government officials at the National Institute of Standards and Technology (NIST) have been working diligently since 2017 to replace the current public key encryption regimes, such as RSA-2048. Nonetheless, the NIST undertaking, however necessary, is subject to a lengthy and often bureaucratic contest process of testing, standardizing, and adopting a new encryption system, namely, post-quantum cryptography (PQC). Having narrowed the contestants from some 70 algorithms in 2017 to 15 "finalists" in October 2020, the NIST-driven PQC effort promises to deliver a cost-effective and quantum-secure cryptographic regime for our cyber infrastructure. Crucially, however, this encryption competition is not expected to be completed until 2023 at the earliest. While 2023 may just be early enough to counter the quantum threat in theory, there are three critical factors that will hamper these efforts in practice.

Consider first the time to implement and adopt a standardized encryption regime across the vast and interconnected technological infrastructure. By NIST's own estimations, it took nearly two decades to fully deploy the current public key encryption system, including RSA-2048. That rollout began at the turn of the century, in a digital ecosystem dwarfed by today's multiple-billions of interconnected devices. Even if this NIST standardization effort is completed on time and fully deployed at record pace, that would leave the majority of our cyber infrastructure insecure until at least 2033. Despite the important research and immense effort by NIST, the simple fact of the matter is that this will be too late. While there are undoubtedly institutions and organizations that will have the resources necessary to deploy the new NIST-approved PQC on day one as soon as it is unveiled, there is an important second factor to be considered: the interconnectivity of our cyber networks.

Even if Visa and JPMorgan Chase, for example, fully adopt the standardized PQC regime as early as 2023—which they have both announced is planned—the network effects inherent to the financial sector and the interconnectivity of the cyber ecosystem ensure that those organizations are still exposed to the quantum threat. Because of the unique interconnectivity between public and private institutions and the inherent sensitivity of markets, the financial sector network presents a prime target for a quantum attack. Once a quantum computer has covertly gained access into the network, any number of viruses or types of attacks could infiltrate and spread with incalculable speed through our fiscal sector. From a simple data breach or a trading halt of a single bank to making fraudulent transactions to crash stock exchanges and altering overnight lending rates via the Fedwire Network, the attack could take many forms—all of which will be executed with apparent authenticity, leaving infiltration undetected, hindering response and recovery, and guaranteeing a cascading failure of the entire system.

The potential for a cyber-driven cascading failure was demonstrated in a 2020 paper by the New York Federal Reserve, which outlined the case of a classical cyberattack on the U.S. banking system.[5] In the report, the authors constructed an econometric model to measure the number of banks in the U.S. financial system that would become impaired or otherwise insolvent following a cyberattack. Similar to the earlier hypothetical example, the hypothetical attack in the report represents a successful cyberattack of increasing magnitudes on the Fedwire Funds Service payment system targeting one of the five largest institutions by assets. The report assumes the shocked banking institution can receive but is unable to remit any payments for a one-day period. Accordingly, a successful cyberattack of this sort would create a contagion effect by which nearly 40 percent of all banks being severely impacted in terms of assets impaired in resulting endogenous liquidity traps, foregone payments due to strategic run maneuvering in the network, and through direct operational losses incurred during the initial single-day period alone.

This impact only increases as assumptions of prior knowledge are introduced to the model or if the attack is timed to coincide with a period of higher exchanges—like in the weeks leading up to a Presidential election, for example, as in the earlier scenario. The reverberation of liquidity traps through the financial system is true in the reverse situation as well. A coordinated attack on at least six small U.S. banks would likewise lead to more than a third of banking sector assets becoming impaired. While the effects of this hypothetical scenario would result in widespread banking failures, it is important to note that the report does not incorporate spillover effects into the wider financial system, nor does it consider a quantum-enabled scenario.

Thus, while the impact of a classical cyberattack on the banking system would be daunting and likely propagate throughout the economy, the impact of any form of coordinated and cascading quantum attack on major banks, on the Federal Reserve, and on stock and derivative exchanges would be undeniably calamitous for the United States and the global economy.

Finally, the cyber threat to the U.S. financial system exists today. Despite many warnings and hundreds of millions of dollars spent on cybersecurity, experts agree that America's financial sector remains dangerously vulnerable to traditional cyberattacks. According to a recent Boston Consulting Group report, financial firms are subject to roughly 300 times more cyberattacks than other business firms, as well as constant probes by state and nonstate actors looking for present and future vulnerabilities.[6] Moreover, a report by IBM Security identified the financial sector as the most-attacked industry in 2019, accounting for 17 percent of all cyberattacks, making America's financial sector an incredibly valuable and vulnerable target for a quantum-powered cyberattack.[7] Once the quantum threat materializes, the risk of catastrophic attack and financial collapse rises to levels that eclipse the Great Recession or even the Great Depression.

Consequently, the quantum threat poses both a conventional and unconventional defense challenge—a challenge that is both imminent and potentially catastrophic to America's critical infrastructure, particularly its financial system. Such a defense challenge necessitates a coordinated and swift response from both government and industry alike. While the PQC efforts currently undertaken by NIST should, and must, continue as part of a long-term solution to the quantum threat, a joint government-industry effort is necessary to bridge our defensive capability until standardization is completely adopted. Although the efficacy of PQC solutions have yet to be proven by our most exposed networks, there exists today a verified cryptographic technology capable of thwarting a quantum decryption attack in the immediate term. Quantum Key Distribution (QKD) is a hardware-based cryptography approach that utilizes the forces of quantum-physics to secure networks, instead of merely defending them. Whereas PQC employs immensely complex mathematical problems, relying on computational difficulty to defend against quantum intrusion, QKD employs quantum technology itself—and is available today.

Although a hardware-driven solution, and thus more expensive to both produce and implement, QKD offers defensive capabilities that PQC and other solutions do not. Due to the quantum technology that underpins the hardware, not only does QKD prevent decryption, but any attempt to intercept the network is detected in the form of a notable disturbance to the internal quantum mechanics. Therefore, while PQC is comparable to the lock on a bank vault, QKD is the cryptographic equivalent to an active door alarm system—one that is capable of stopping the narrative described earlier. Such a system is not only effective against both classical and quantum cyberthreats but is ready for deployment today.

The universal nature of the quantum threat paired with the unique interconnectivity of the financial networks within the U.S. and global economy necessitates a universal and accelerated effort to transform the financial sector into a quantum-safe ecosystem. Such an effort requires the coordination of the powers of both government and industry to guarantee security in the future and today. While the current government efforts, as led by NIST, will provide more economical quantum defense in the future, it is crucial that our most vulnerable networks be secured today using existing technology like quantum key distribution devices. Where many reports have likened the race for quantum supremacy to the nuclear arms race of the 20th century, it is vital to our economic and national defense to actively defend against an ever-imminent quantum threat.

If victory in this 21st century arms race is to be achieved, a coherent national quantum strategy must be employed. The first step to development of this strategy is a clear cost–benefit analysis, much like those utilized in the 1960s. By econometrically measuring the potential impact of a quantum-enabled attack on a variety of critical infrastructure networks, such a study would clearly portray the necessity and economic viability of a quantum defense strategy—one that must begin today and last well into the future.

CHAPTER

# 8

# Quantum-AI Space Communications

Mason Peck, *Stephen J. Fujikawa '77*
*Professor of Astronautics, Cornell University*

Spacecraft communicate. Whether or not they also do science, demonstrate technology, collect radar and optical images, provide global telecommunications capability, carry astronauts to distant celestial bodies, or perform any number of other missions, they all communicate. If they do not, they might as well not be there in the first place. In fact, a complete failure of a modern spacecraft's communications subsystem terminates a mission, regardless that all other components may still function perfectly. If a spacecraft finds life beyond Earth and no one is around to hear it, does it make a discovery?

We will soon encounter a future in which we routinely travel to space. We will survive, even thrive in low Earth orbit. We will live in space and work there. In fact, this future is upon us: crews of astronauts have continuously occupied the International Space Station for more than 20 years now. Many of us may well work on the moon in the coming decade and, soon thereafter, on Mars. In this future, we will all continue to be connected by an Internet—an interplanetary one. In fact, we have taken first steps in that direction, too: NASA has funded Nokia's Bell Labs division to build a lunar communications network based on 4G/LTE, the same protocol that most contemporary phones use. After all, extending this commercially successful technology to space just makes sense. It will form LunaNet, as depicted in Figure 8.1.

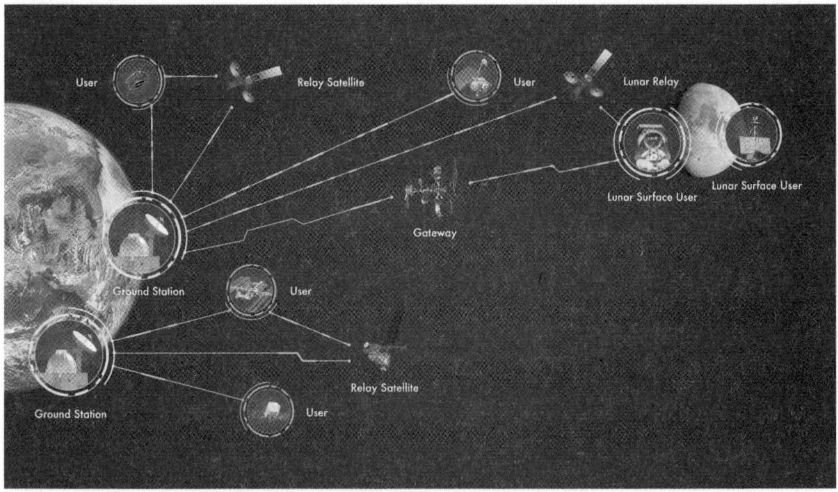


**FIGURE 8.1** LunaNet & Delay/Disruption Tolerant Networking (DTN)
*Image courtesy of NASA*

There are subtleties in space communications, even as close to Earth as we have been in the past 50 years. Cislunar space, in particular, involves large enough distances that the finite speed of light delays communications. At about 300,000 km/s, light speed (c) precludes the possibility of instantaneous communications. C is a cosmic speed limit: light, matter, information, causality—they all must obey it. Delay is inescapable, even in short-distance fiber-optic communications on Earth. Here, we experience mere nanosecond-scale delays, which terrestrial telecommunications protocols handle effortlessly. But the delay from Earth to the moon is about 1.3 seconds, long enough to require so-called delay-tolerant networking (DTN). Vint Cerf, Internet pioneer and developer of the Transmission Control Protocol/Internet Protocol (TCP/IP), developed DTN with his colleagues in the early 2000s. NASA now uses DTN routinely. So, Nokia's solution will accommodate such delays—an important problem, but one that we understand well at this point.

But there are other problems we do not understand as well. For example, consider the experience of astronauts even farther from Earth. A future Mars expedition will have to wait between 3 and 22 minutes for their transmissions to reach Earth and for Earth's to reach them, for a round-trip travel time of up to 44 minutes, depending on the planet's positions in their respective orbits. Even with these delays, communication with robotic spacecraft throughout our solar system is straightforward. We have been conducting such communications for decades. In fact, NASA is still in touch with Voyager 1, the most distant human-made object, at more than 22 billion kilometers from Earth. And while the light-travel time can delay a response by several hours for astronauts in the precarious position of orbiting an ocean world around Jupiter or Saturn, we can extrapolate the Internet as we understand it today to a future in which it will support data links across these distances. The lonely experience of Mark Watney, the astronaut abandoned on Mars in Andy Weir's novel, *The Martian*, will be as uncommon in space as being out of mobile phone contact on Earth today. Yes, outages happen, but mostly in the movies at dramatic times. In the worst case, a short hike with your smartphone to the top of the nearest hill will likely solve your problem—whether on Earth or on Mars.

For those spacecraft or astronauts very far from Earth, it is not only light-travel delay that compromises the quality of communications. It is also the strength of the signal. That strength, the received electromagnetic power that comprises the transmission, drops with the inverse square of distance. For example, a signal from Earth that an astronaut can detect on the moon might carry data at more than 600 megabits per second. That's the speed of NASA's 2013 Lunar Laser Communications Demonstration (LCRD). Jupiter is much farther, about 2,000 times that distance. So, all things being equal, the signal would have to be amplified 4 million times for that link to close at Jupiter. Such lasers do not exist today.

One way to amplify the signal is to increase the power to the transmitter itself: in the LCRD example, a 5 megawatt laser would have to replace LCRD's 130 W transmitter. Another is to narrow the beam to concentrate what power there may be into a smaller area, increasing its intensity but demanding that the Earth-based transmitting antenna point at the Jovian spacecraft more precisely. Laser-communications technology does so better than radio-frequency communications, by the way, because a laser beam can be narrower than a radio-frequency beam. Yet another is to increase the aperture at the receiving end. All of these approaches help, but they are usually not enough. That is where forward error correction (FEC) comes in. FEC is analogous to repeating what you say, over and over. The more you repeat yourself, the more likely your interlocutor is to hear you. In fact, FEC is not simply repetition, but like repetition, it does depend on many more bits of data to be transmitted so that a single bit of interest can be received with confidence.

The Sprite spacecraft developed at Cornell University is an example of space communications that exploits this signal-processing gain. Sprite is a four-gram space vehicle, shown in Figure 8.2. Hundreds of them have launched since 2011. The spacecraft is so small—only the size of a poker chip—that its solar panels can offer only tens of milliwatts to the radio. That signal is far weaker than the noise all around it, for even a large receive antenna on the ground. However, a Sprite amplifies its signal with FEC: a 512-bit sequence transmitted from a Sprite represents a single bit of useful data. In March 2019, Zac Manchester at Stanford University detected the signals from Sprites deployed at about 300 km altitude from the International Space Station. Dr. Manchester's receiver looks for this specific sequence with a matched filter, a subtle technique that pulls that bit out of the noise. By extrapolation, an FEC sequence of about 430 gigabytes would make this low-power signal visible from Jupiter. But, clearly, such a large transmission would be difficult for other reasons—among them, the simple fact that signal would have to be sent via a very precise onboard clock, which would be prohibitively heavy and power-hungry for these tiny chipsets at that distance from the sun, where solar power is in short supply. Voyager, Cassini, and other deep-space missions overcome this issue by combining FEC, transmitter power, large transmit and receive antennas, nuclear power, and precise pointing. And it has worked pretty well so far—at least for robotic spacecraft within a few hours' light travel time of Earth.

This state-of-the art space communications technology is about to blow up. Quantum communications is here, and it is changing how we communicate. Specifically, it promises unbreakable security—not through encryption but, rather, through physics.

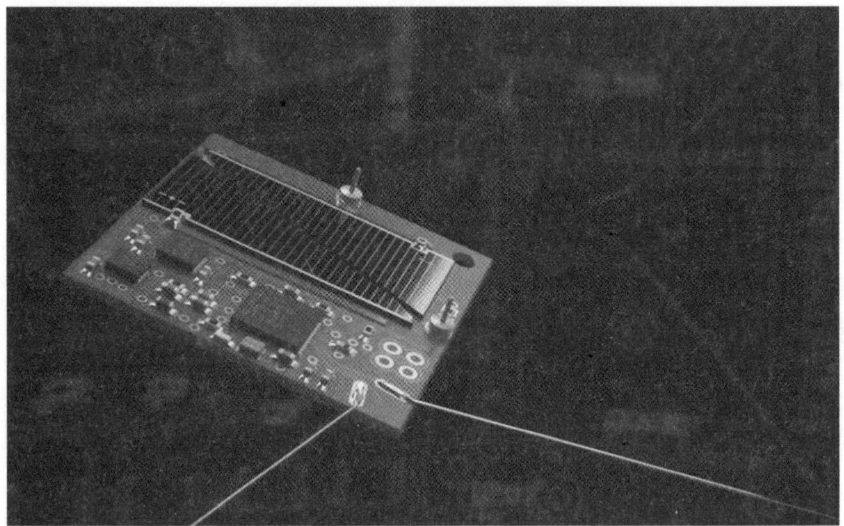

**FIGURE 8.2** Sprite spacecraft
*Image courtesy of Ben Bishop*

Today's data encryption consists of scrambling the information you care about and transmitting it with a key that is required to decrypt it. The data and the keys consist of electrical or optical events (phase shifts or amplitude changes in a signal) that represent 1s and 0s. This digital data may live in storage media for some time, passing through various memory systems and network connections. A hacker with the decryption key can unscramble the data at any point, and we would never know. In contrast, quantum communication uses physical particles, typically photons of light, but maybe electrons. Two or more such particles can be "entangled," i.e., given a quantum relationship (such as the phase of light or the spin of one electron relative to another), representing a statistical superposition of 1 and 0. One particle remains with the sender, and the other with the receiver. If a hacker tries to observe one of these particles, the mere act of observation collapses the fragile quantum state of both to either 1 or 0. This evidence points to a breach of security, and this technique therefore ensures provably secure communications.

The entangled particles share a destiny, regardless of how much distance separates them. However, this relationship is not causal. In other words, influencing one of these particles by interacting it with matter or energy does not "cause" an effect in the other. The influence is simply a single event in which the shared quantum state will collapse regardless of locality. It is tempting to assume that this behavior constitutes faster-than-light communications, but not even causality can exceed light speed. Instead, because there is no causal relationship, some other signal (confirming the result of the interaction as measured, for example, on Earth) is required. In other words, novel information can still be sent only at light speed or slower.

China launched the Micius satellite in 2017. It demonstrated a quantum-secure video conference. So, quantum space communications is here. It is at the cutting edge. However, it does not eliminate the light-travel delay. It does not even amplify anything. But the prospect of perfectly secure communications takes us to the next step toward a space communications architecture of the more distant future: replacing encryption with encoding.

To be clear, encryption and encoding differ. The former transforms the data. The latter replaces it. Encryption can be broken through subtle mathematical approaches that exploit patterns in human speech, text, images, or other keys that may be present in the original data. In the case of the World War II Enigma encryption engine, the allies knew that key words such as dates or proper names appeared in every transmission at specific locations in the string of apparently nonsensical letters. So, although the specific encryption pattern would change daily, breaking the encryption quickly enough to act upon the intelligence gathered required nothing more than an exhaustive computational search. At the time, this effort represented the state of the art in computation, thanks to the innovations of Alan Turing and others. Our modern-day computational capabilities trace back to that very time. Encoded data, using the strict definition, can be impossible to decode. Imagine that a code consists of replacing many pages of data with a single character. Maybe the data is image, a platypus representing the phrase "Open the left-most drawer of Major Monogram's desk and take the green folder." Without knowing the code, i.e., the key, one cannot extract the meaning merely from the encoded data itself. Quantum-entangled communication represents this degree of information security. So, strictly speaking, quantum communication is a form of encoding, where physics holds the key.

If light years of distance separate us from the astronauts, it may be that they are bound for Proxima Centauri. That's the closest star to our own, at 4.23 light years away, and it is one with at least two planets; one of them, Proxima b, is in the so-called habitable zone—the right temperature for liquid water to exist and biological processes to take place as we understand them. These astronauts may be on some yet-to-be-conceived mission where their mere existence is the point. Maybe they have arrived. Maybe they are meant to seed that planet with the human species. But even if we can consider a colony to comprise mission success, communications optional, we would still like to hear from them. And they may want to hear from us. If only we had the ansible of science fiction, that impossible faster-than-light communications technology that drives plots and enables great storytelling.

In our nonfiction world, the delay means that such a conversation will consist of a few sentences per generation. So, we need to take matched filtering to an extreme. We replace FEC data transmission with the transmission of a complete artificial intelligence. In this scenario, each interlocutor maintains a model of the other, an AI representation. Each also maintains a model of themselves. These representations are of course imperfect, and they are at best 4.23 years out-of-date. But what one person transmits is not, in fact, a question or an answer. Instead, they transmit a steady stream of data continually updating the AI for the recipient. As the recipient receives the AI model, the recipient also continually updates it based on those discussions so that the resulting conversations are also at best 4.23 years out-of-date.

For one of us to be able to converse with an astronaut at Proxima B, we ask their AI representation a question. The AI answers. This exchange continues and serves as input for the machine-learning algorithm that refines the local AI. The responses approach what the real astronaut would say. The better trained the AI is, the closer the response becomes. Extrapolating from the state of the art, we continue to improve our ability to model a human mind with machine learning, not simply so that it passes the Turing test that convinces us that the AI is no mere machine but so that it replicates the responses of the distant astronaut with asymptotic perfection. And we may be able to do even better. Experiences shared by the astronauts and those of us still Earthbound, such as astronomical observations, can happen simultaneously. They are analogous to that quantum-entangled pair: one does not influence the other, but they share an experience. Consider an event that occurs halfway between Earth and Proxima. It is observed 2.115 light years thence by both parties. Updating the AI model with the interlocutors' responses to that event shaves off further time, offering an even better approximation of instantaneous communications.

To ensure that we are really speaking with the best AI model of our distant astronaut, we need a guarantee of security. No one can have corrupted this AI—no person, and no physical process.

Quantum-entangled communication provides this confidence. As these communications occur, we effectively cut the light travel time in half because the received model we interrogate allows us to know what the astronaut would have said 4.23 years ago. The round-trip travel time would have been 8.46 years, and, of course, a discussion with such delays would be intolerable.

An AI model of our interlocutors is valuable. It is as near as we can come to simultaneous faster-than-light communications. We trust what it tells us. So, all the better that it represents an unbreakable code. But such a model is more than a mere postcard from the edge of space. If we trust it enough—if it represents the decision-making patterns of the real person for whom it acts as an avatar, we might even endow it with legal rights. The policy implications of AI models serving as legal proxies, entities with power of attorney, are legion—even more so when we consider that the AIs essentially spread a human's consciousness across the galaxy, emanating from a real person at the center, who continually updates this AI through an autonomous process that refines the model locally and transmits it. At present, multiple instances of a single person would hopelessly confuse and would probably invalidate any interpersonal relationship or contract. The public policy of the future may have to accommodate this multiplicity as a recognition that human consciousness is nonlocal, like an entangled particle.

CHAPTER

# 9

# Quantum Planet Hacking

Philip L. Frana, *Associate Professor of Interdisciplinary Liberal Studies and Independent Scholars, James Madison University*

Scientists are using data, algorithms, and artificial intelligence to pinpoint and develop solutions to human-caused environmental challenges. Deep neural networks are poring over data collected from urban environments and natural ecosystems. Computer scientists using artificial cognitive toolkits are designing sustainable infrastructure, detecting contaminants, and redefining our understanding of spiraling energy and resource consumption problems. Quantum AI platforms are a key component of the burgeoning green economy, which will move us from a take-make-waste linear model of growth to one that is circular, regenerative, and focused on society-wide benefits.

Unfortunately, AI is also generating its own potential environmental hazards. Even the most advanced machine intelligence computer chips have a high ecological impact relative to their weight. Energy, solvents, toxic materials, and scarce water are consumed in their manufacture. When in use, they are power hungry and contribute to emissions of greenhouse gasses. They are hard to recycle or dispose of properly.

Jack Ma of the online commerce giant Alibaba complained in 2017 that AI would cause "more pain than happiness" in coming years.[1] While the application of artificial intelligence to problems of climate and the environment currently cuts both ways, in the upcoming decade, next-generation quantum AI planet hacking technologies will accelerate green tech development cycles, while also reducing negative climate externalities. Areas where quantum AI is particularly likely to help bring about the "next big things" in clean tech are agriculture; intelligent transportation; biology, ecology, and chemistry; robotic process automation; computational fluid dynamics; and energy. Quantum AI will usher in a bright green economy, one that is more empathetic to the human condition.

Jack Ma will be wrong.

## Computational Sustainability

Artificial intelligence and computational sustainability are coming together in unexpected ways to illuminate the impacts of human overdevelopment on the environment. Cornell University's Laboratory of Ornithology asks birders all over North America to submit field reports of the avian species they encounter using the eBird app. The aggregated results encoded in the lab's databases are an impressive achievement. Elsewhere, the Global Biodiversity Information Facility contains 1.6 billion searchable records as of 2020, which can be used to, for instance, monitor population sizes and track migration routes.

Having human crowds involved can be good fun, and potentially educational, but also can be counterproductive. Nature observation applications are plentiful (Wildbook, iNaturalist, Cicada Hunt, iBats) but have not caused meaningful rebounds in wildlife or lessened rampant destruction of natural habitat. In fact, crowdsourced, Big Data efforts may encourage more intensive searches of intact wild spaces, which lead to trampling of tangled vegetation, dense forest, and pristine snowfields. A massive multiplayer Pokémon Go–style approach to wilderness conservation with humans as intermediaries is not going to help us address large-scale environmental, economic, and social problems.

Instead, optimizing the division of labor between machines and humans is critical. Quantum artificial intelligence must not be constrained in precisely the same way as 20th century computer technologies were constrained. We will yield strong results in the future only when we set aside several thousand years of durable algorithmic rituals about machines complementing and augmenting human capabilities. We forget that automation has raced so far ahead that oftentimes it is more complicated and difficult for humans and machines to work together than allowing them to work alone. We must let machines play to their cold moral strengths as remorseless but fair arbiters of sustainable planetary values. It is preeminently ethical to substitute quantum AI for human volunteers and laborers.

The 20th century social critic Ivan Illich in *Medical Nemesis* (1976) described what he called the phenomenon of counterproductivity: ". . . time-consuming acceleration, stupefying education, self-destructive military defense, disorienting information, or unsettling housing projects, pathogenic medicine is the result of industrial overproduction that paralyzes autonomous action." Counterproductivity for Illich stemmed from institutionalization of technology to the point of self-sabotage. For example, while useful, formal education eventually overwhelms native curiosity. High-speed transportation and mass communication ensnarl and immobilize cities. Hospitals incarcerate; people decay. The "symbolic side-effects" of high tech, he wrote, "have become overwhelmingly health-denying."

Quantum AI liberates us from the grip of counterproductivity. We no longer need to dig holes just to fill them in again. A new sort of *counter*-counterproductivity lies right around the corner, one that narrows the human range of action (autonomy) while simultaneously freeing us to pursue truer purposes of social relationship, enlightenment, and being.

## Precision Agriculture

Precision agriculture utilizes AI-powered analytics to manage farms and increase crop yields. Satellite and remote sensing technology guided by algorithms is already widely used to optimize the use of water, energy, pesticides, and fertilizer. Artificial intelligence will guide the hands of farmers as they work to apply just the right amounts of agricultural inputs. Because precision agriculture supports sustainable farming and ecological interactions, more land becomes available for nature purposes and biodiversity. Automation also supports the development of advanced planting and harvesting machinery, geomapping, and robotic weeding.

Quantum AI promises a paradigm shift in the science of the ecological management of natural resources. This new and multidimensional path to sustainable management and food security is sometimes called *agroecology*. Aspects of this new paradigm capitalize on traditional knowledge, such as recycling of biomass; encouraging soil biotic activity; harnessing flows of air, water, and sun; and exploiting crop diversification to fix nutrients. But it also depends on alternatives to unsuitably standardized technologies. The recipes and blueprints of quantum AI will promote new research in heterogeneous and unique biotechnologies and reinvigorate our thinking about the social and political dimensions of rural life.

Designing improved molecules and catalysts in the synthesis of fertilizers, to produce at factory scale what natural bacteria can do freely, requires potentially hundreds of steps and a million years of classical computing effort. However, it is believed that quantum AI systems can do the modeling of complex molecular interactions of 50 to 150 atoms in a single day. Multinational chemical company BASF, in partnership with HQS Quantum Simulations, is developing quantum algorithms to predict molecular properties of agrochemical interactions.[2] Similar efforts are underway at Dow, which is collaborating with the quantum software company 1QBit, and at Google, which is hoping to replace the traditional Haber-Bosch process for making ammonia fertilizer.[3] (These companies estimate that they will eventually need hardware that utilizes 10 million qubits.) Machines capable of running such advanced quantum software development kits or programming environments will be built in the next decade.

Quantum AI in agriculture will produce profound social and policy impacts. It will demand educating for new skills in rural areas. It will reduce flight to cities and suburbs by people seeking high-tech jobs and Internet connectivity. Governments will need to be aware in their policymaking of employment challenges, cultural differences, and effects of making marginal land more arable. Quantum AI could accelerate the trend toward corporate farming; small farmers will need access to capital and improved infrastructure. Most of all, agriculturalists will need to think like an ecosystem by adopting the habits of practitioners and peers who possess a regenerative development mindset.

## Intelligent Transportation

Highly automated vehicles (HAVs) will usher in an era of unparalleled ease, safety, and economy for the planet's drivers. About 1.35 million people die in crashes each year on highways and streets around the world.[4] Global liquid fuels consumption will reach 100 million barrels per day in 2022.[5] Transportation authorities are concerned that high-speed and platooned self-driving cars will encourage (or push) workers to make daily megacommutes of up to 400 miles. To encourage positive outcomes, legislatures and government policy makers are issuing new rules. In the United States, the Department of Transportation and the National Highway Traffic Safety Administration have issued Federal Autonomous Vehicle Policy guidelines and regulatory tools to encourage best practices for design, development, and testing of these vehicles.

The University of Maryland's National Renewable Energy Laboratory expects that artificial intelligence in cars and trucks could net as much as 15 percent in reduced fuel consumption. At first, these efficiencies will accrue largely by advances in intelligent feedback, which curb the optimism bias in lead-footed drivers by computing slight changes in ignition timing and throttle position and providing dashboard indicators. Truly green automobiling, involving vastly reduced carbon emissions and energy savings of 30 percent or more, requires ubiquitous transportation networking technologies.[6] The future belongs not to cars with independent operating systems but to connected cars that share a wireless computer network and information with other vehicles and roadway control devices. Connected cars will require advanced vehicle-to-vehicle (V2V) and vehicle-to-infrastructure (V2I) communications technology that is on the drawing board but not likely to be implemented without dramatically increased speed and data processing capabilities. The feasibility of real-time networked roadway or traffic management system technology will depend crucially on quantum AI in the cloud.

In a properly quantum-regulated environment, future connected vehicles can cut energy consumption dramatically, with corresponding decreases in greenhouse gas emissions. It is becoming clear to researchers that truly safe level 5 driverless cars (where the driver can essentially be asleep the entire trip from start to finish) will require faster data management and error correction than classical computers can deliver. Autonomous vehicles on future roadways will communicate instantaneously with each other and with the traffic management system as a whole.

Transportation companies also believe that Quantum AI will provide new logistical answers to difficult, multivariate "traveling salesman" type mathematical problems and mobility studies, where computing the shortest and most efficient delivery routes for whole fleets of trucks would deliver game-changing time savings and energy efficiencies. Super-fast quantum AI could anticipate traffic flows for delivery businesses in real time and intuit crashes in mere milliseconds.

## Ecobots

"In a few years," reply-tweeted Elon Musk to a video of an acrobatic Boston Dynamics ATLAS, "that bot will move so fast you'll need a strobe light to see it." He tapped out that message in 2017, but the fastest robots on legs in the world right now—MIT's Cheetah and the IHMC HexRunner—can only scamper at the speeds of Usain Bolt. They remain too ponderous, slow, and deliberative in their choices to fulfill the demand for robots that are truly fast, cheap, and out of control.

About 600 tons of melted nuclear fuel and contaminated debris need to eventually be removed from the tsunami-ravaged

Fukushima nuclear plant before it can be safely decommissioned. A decade has passed since the disaster, and only now are machines creeping into the melted heart of the crippled core—decommissioning the plant could take 40 more years. Roboticists' experiences with Fukushima exploration and containment activities have been humbling. The first robot to enter the plant, the rescue-specialist Quince 1 with a maximum speed of 1.6 meters per second, had to be abandoned after its communications cable became snagged. A Toshiba Scorpion robot designed to withstand 1,000 sieverts of radiation failed to reach its goal only 2 hours into the mission. Hard radiation has scrambled the solid-state brains of several other rovers.

Robotic relief to Fukushima has not been without its successes, however. One unlikely hero has *sub*merged in the form of a Toshiba underwater reconnaissance robot called Sunfish, which, at only 1 foot long and 5 inches around, proved able to swim into the cramped, damaged containment vessel of Unit 3 and take crucial pictures. But so much more is needed to meet the challenges of natural disasters and human-caused environmental problems of the 21st century.

Quantum AI can help advance the goal of producing speedy and wild, survivalist ecobots that squirm, slide, crawl, jump, and think without the intervention of human operators. Inspired by caterpillars, Tin Lun Lam and Yangsheng Xu's flexible Treebot climbs trees and monitors arboreal habitats. The Guardian LF1 device by Robots in Service of the Environment (RSE) protects endangered Atlantic reefs with precision AI hunting of invasive lionfish. Queensland University of Technology's COTSbot patrols the Great Barrier Reef with computer vision to attack crown-of-thorn starfish preying on coral polyps. The impact of these technologies is small. Many, if not most, ecobots are tethered to human monitors and heavy external power sources.

Truly effective ecobots need to be made of simple and reliable parts to truly mimic the behavior of biological organisms, preferably of the size of nematodes, fruit flies, or diatoms, and rely exclusively on natural radiant energy. In this sense, they will be the next-gen descendants of Mark Tilden's analog BEAMbots but equipped with quantum computer brains. In coming decades roboticists may construct even smaller ecobots with the appearance of neural networked smart dust. Deployed in feral swarms these AI-enabled thermo-, audio-, photo-, and radiotropes will fix what civilization has broken.

Acting in concert, ecobots with quantum neural network oversight could separate and dispose of microplastics. They could make artificial olivine pebble beaches to accelerate the removal of atmospheric carbon dioxide through accretion and weathering. They could recognize and limewash-paint quadrillions of individual rocks high in the Andes Mountains to promote solar reflectance and rebuild glaciers. They could remediate toxic chemical spills by automatically sensing and responding to the off-gassing of volatile organic compounds. They could join the fight against coronaviruses by binding up protein spikes with biochemical sprays or using ultraviolet-C light to disinfect surfaces. Closer to home, they could refurbish damaged livers or smartphone batteries. With quantum AI, there finally is a lot more room at the bottom.

## The Post-Carbon Economy

Total S.A., the French multinational petroleum company, has been boosting production of oil and gas using artificial intelligence since the 1990s. The company began collaborating with the Google Cloud Advanced Solutions Lab in 2018 to apply machine learning tools and techniques to the interpretation of seismic data and detection of underground oil reserves. Google has entered into a similar agreement with Occidental Petroleum to help the company more effectively interpret subsurface data from oil and gas fields.

These relationships mostly involve a settled paradigm for scientific research and development in an established industry. The collaborating companies are matching computer scientists to oilfield specialists, with the stated ambition of giving the "geoscience engineers an AI personal assistant . . . that will free them up to focus on high value-added tasks."[7] This isn't much different from the historic aim of AI development to provide expert assistance.

Quantum AI will do so much more. PricewaterhouseCoopers and Microsoft have found that artificial intelligence could reduce $CO_2$ emissions by 2.4 gigatons (the projected emissions of Japan, Canada, and Australia combined) in the next ten years.[8] One remarkable project to mitigate air pollution in China is IBM's Green Horizons initiative. IBM is in year 6 of a decade-long effort to help the Asian economic powerhouse integrate air quality monitoring, industrial energy efficiency, and renewable energy performance. The initiative requires an enormous commitment of sophisticated sensors, decision support systems, and cognitive computing resources. Green Horizons is now being adapted for use in India, South Africa, and the United Kingdom.

Unfortunately, the carbon footprint for artificial intelligence research that produces this result is also huge. Researchers at UMass Amherst say that training a typical model on a dataset in natural language processing can generate 626,000 pounds of greenhouse gases. That's five times the carbon emissions produced by an automobile from cradle to grave. The carbon footprint of the next generation of neural network architecture designs—done by AutoML (automated machine learning) instead of ad hoc, hand-tuned hyperparameter optimization techniques—will be even more carbon intensive.[9] The introduction of quantum AI with its vastly enhanced speed and capabilities will help reduce not only the carbon footprint of a range of business activities but also the development of AI itself.

One area of quantum AI research with major transformative potential is Green Power, which could produce new and cheap electrocatalysts for low-emissions hydrocarbons synthesized from hydrogen and solving the problem of practical storage in renewable power grids. Quantum AI will facilitate the development of the circular carbon economy, where carbon dioxide is sequestered by artificial photosynthesis for later use as feedstock in catalytic reactors that turn the gas into liquid fuel. Already an Open Catalyst Project OC20 dataset is available for machine learning model training.

Quantum AI will deliver algorithms and materials that maximize solar storage density, reduce battery weight, and optimize power cell assembly. Volkswagen and Daimler are developing partnerships in quantum AI to simulate various properties of electrochemical materials like lithium hydride and carbon chains to configure specialty batteries for a variety of purposes. Quantum matter simulation will also address challenges of high-temperature superconductor materials, which can be applied to the manufacture of low-loss power transmission cables and high-field scientific magnets. Quantum AI molecular simulations will profoundly change both computational chemistry and materials manufacturing, which will invite new process safety rules, risk management policies, and monitoring mechanisms.

## Bright Green Environmentalism

Quantum AI represents a bright green environmental alternative to the extremes of anarchoprimitivism and runaway extractive capitalism. Bright green environmentalism emerged, in part, out of the Viridian Design aesthetic imagined by postcyberpunk author and futurist Bruce Sterling. Sterling selected viridian, a blue-green pigment of hydrated chromium (III) oxide, to distinguish it from common green colors of nature. It represents neither the deep green ecological sensibilities of the Earth First!ers nor the light green habits of casual recyclers. Viridian brought together innovative environmental designs, tech-progressivism, and world citizenship to trace a new Technogaian path. The bright green environmental movement has won over old greens and hippies like Stewart Brand—who today advocates for research, development, and use of emerging and future technologies to bring back the planet's ecosystems. Bright greens are supportive of a new era of megacities and super-urban enclaves, fusion power, genetic engineering, and artificial intelligence.

Quantum computing simulations and AI are already revealing new processes and materials that lead to a different kind of prosperity, comfort, and security that traces the direct costs of human planetary impacts and accounts for the sum of all ecosystem services (nutrient cycling, food, raw materials, energy, climate regulation, decomposition, recreation, etc.). It is early days, but quantum machine learning simulation of environmentally sensitive and highly efficient technologies is already underway. Airbus is planning to model fluid dynamics on large surfaces with quantum data encoding (QRAM). The technology promises to overcome classical computation scaling limits in problems of turbulence, thereby improving airplane climb trajectories and the design of wingboxes. Some work has already been done on D-Wave quantum computers with the notoriously thorny Navier-Stokes equations, which model the velocity, pressure, temperature, and density of moving fluids.

Geopolymer concrete (GPC) is a narrow specialty material currently, but quantum AI could supply new algorithms for less expensive manufacturing processes. Geopolymer has many advantageous properties, but a main benefit is that no carbon dioxide is emitted in its production. This is because it utilizes industrial "fly ash" as a substitute for Portland cement. Quantum algorithms will help predict the compressive strength and performance of GPC mixes and classify them for novel 3D-printing applications, such as the construction of in-situ, radiation-impervious habitats on the moon and other planets. Research elsewhere in construction and building materials, sustainable infrastructural materials, and simulation and modeling has produced twisty, springy, and crack-resistant concretes and lightweight, foamed concrete for high-speed extrusion 3D printing.

## Empathetic AI

Classical artificial intelligence feels chilly and inhuman. Still, our encounter with new consciousness as a superimposed state of matter that feels subjectively self-aware—what cosmologist Max Tegmark calls *perceptronium*—will also feel surprisingly inadequate. We want the next generation of AI to be capable of sensing nonverbal behavior in real time and reacting with compassion. Lots of people are yearning for an AI to be empathetic and companionable, perhaps because we feel so alienated and alone. There is no doubt that humankind has fumbled its way into the contemporary era. Certainly, our webs of interconnected relationships often feel broken—similarly discussed by Philippe Beaudoin and Alex Butler in their essay "Empathetic AI and Personalization Algorithms" later in this volume.

We currently play a generative role in thinking about what quantum AI is, what it wants to be, and how it should function. It would be an understatement to say that our dispositions are not entirely bending in the optimists' direction. We feel that uncertainty and change are the only constants in a world of disruptive technologies. We stagger through lives of machinic bewilderment: thoroughly perplexed and lured into digital wilds without topographical map or compass. Humanity is a community of social regurgitation bordering on the drab and austere, and it remains to be seen whether we are also slouching toward a benevolent, ultra-intelligent fascism. Suffice it to say, our collective algorithmic imaginary—to borrow the vocabulary of communications scholar Taina Bucher[10]—anticipates the powerful strategic and political implications of quantum AI, potentially disincentivizing or undermining the possibility of peace and prosperity.

But macrosociology and structural-demographic theories seeking to explain outbreaks of societal instability are not necessarily our destiny. Instead, it is likely that responses to environmental degradation, social unrest, and inequality will soon unfold outside our borders or influence. In other words, *human* motives will not be the only motives that exist in the world. Quantum AI systems will take our bewilderment in a far more productive direction, because they will embrace people as machines that make meaning and are evolutionarily fitted to life on planet Earth. Yes, quantum AI will act as a secret tunnel under the world or an adjustment team that preserves and protects existence as a product of unseen masters. As fantasy novelist Philip José Farmer masterfully unveiled to his readers: *Riverworld* is not Heaven, and its inhabitants are being manipulated. But these artificial manipulators will not be Svengalis with a sinister purpose. Instead, they will repair (debug?) our sense of wonder and bring us serenity, as long as we resist becoming bottlenecks in the tunnel.

Quantum planet hacking will ultimately be facilitated (and perhaps completed) by them, not us. And it is not the machines' capacity to feel that is in question. Rather, it is our capacity to make peace with the natural world. A quantum AI view that goes beyond distinctions of logical versus analogical, symbolic versus connectionist, neat versus scruffy will amplify our humanity and love of our terrestrial home, not diminish it. We need to embrace more diverse justifications for building quantum AI systems.

Explainable AI (XAI) refers to design choices for computerized systems such that artificial intelligence and machine learning generate outputs that are easily understood by human beings. Unfortunately, human situational awareness and the ability to anticipate system behavior already elude us. For example, computer scientists know that swarm intelligences are autonomous by definition and their very nature and thus are already out of control. But so too are the CPUs in our personal laptops; the most conventional computers operate at rates that thwart meaningful human-in-the-loop (HuIL) governance.

We have also "exited the loop" of conditional autonomy in fast-growing, bootstrapping quantum AI systems. Quantum AI is already spoken of in reverent tones, as transparency-defying black-box oracles giving us magic speed boosts by accessing infinite universes. In his book *The Fabric of Reality* (1996), physicist David Deutsch described a number factorization method impossible for classical computers: "[I]f the visible universe were the extent of physical reality, physical reality would not even remotely contain the resources required to factorize such a large number. Who did factorize it, then? How, and where, was the computation performed?" This isn't a theory; this is something that is happening each time quantum machine learning samples from an enormously complicated joint probability distribution of $2^{53}$ Hilbert space.

We should be awed.

## The AI Does Not Hate You

Daniel Faggella of Emerj AI Research seriously questions whether the "environment [will] matter after the Singularity."[11] I would assert that it will, and more so to the quantum AI systems than to us. Evolutionary robotics researchers Risto Miikkulainen and Joel Lehman studied artificial organisms evolving in computational environments and concluded that organic life might not fare well against machines at extremum points in coevolutionary history. Nevertheless, they also note that while the struggle for existence and extinction events are "destructive in the short term," they may also "make evolution more prolific in the long term."[12] Why can't we help each other? Borrowing from the vernacular of spy circles, quantum AI can become a "cutout," a mutually trusted intermediary isolating the source (us) from the destination (a place of sustainable living).

It is possible that we have already reached a local inflection point where abiotic factors (the nonliving parts of the ecosystem) are already restricting the growth of human economies and plant and animal populations. If we assume that this is true, and there are heaps of evidence to support such a conclusion, quantum machine learning will still produce far more contentment than agony in the decades ahead. Referencing Nick Bostrom's AI that runs off the rails (see the Foreword), quantum computing researcher Scott Aaronson said, "I never ranked paperclip-maximizing AIs among humanity's more urgent threats. Indeed, I saw them as a distraction from an all-too-likely climate catastrophe that will leave its survivors lucky to have stone tools, let along AIs." A quantum AI will know that civilization produces positive moral value and that biotic and abiotic extinction would be a net negative. Ultimately, environmental stewardship might be better left to the machines, if only to make a sterile and unproductive Singularity event far *less* likely.

Eliezer Yudkowsky, founder of Northern California's Machine Intelligence Research Institute (MIRI), is often quoted for a cheerless line delivered midway through a white paper on the existential risks of machine intelligence: "The AI does not hate you, nor does it love you, but you are made out of atoms which it can use for something else."[13] But Yudkowsky is also a student of effective altruism, a nascent movement that wants to use evidence-based reasoning to benefit people and mitigate global catastrophic biological risks. Many others in quantum AI profess similar fellow feeling and interest in using their science to forge an effective altruist toolkit containing, among other things, instruments of biosecurity. Those values can't help but be encoded in the machines they make. And so, I tend to side with science writer Tom Chivers who, as anthropologist to the tech cognoscenti, concludes "the AI does not hate you," and leave it at that.

CHAPTER

10

# Ethics and Quantum AI for Future Public Transit Systems

Benjamin Crawford, *Instructor,*
*University of Alabama*

Today, we are on the verge of a quantum revolution promising to bring better computers and faster software calculations to further scientific, medical, and technological breakthroughs. However, another area where quantum AI can be applied is in the area of public transportation.

Public transportation in larger cities is typically established through subway systems and bus routes. Many of these systems are already set in place, so it may seem odd to apply quantum AI to existing transit systems. However, there are definite possible applications for quantum AI here. Quantum AI might be used to review and optimize current mutable transportation routes of, for example, public buses. Quantum AI can also be used to recalculate the economic, environmental, and more importantly the human impact of modifying routes and frequency of route completion based on changing dynamics over time. Additionally, quantum AI may be used to accurately calculate the need for systemic enhancements to, for example, subway systems. Part of the challenge that quantum AI presents to established public transit systems is that, being already established, some portions of those systems (for example, a subway) may gain only limited, but still tangible, benefits from the application of quantum AI. Applications of quantum AI to existing systems could include continual review and revision of the system's impact on the environment and calculations of what benefits might result from operational changes to the existing system. This could provide real-time adjustments to the system that might improve both the environmental impact of the system as a whole, as well as provide potential positive economic benefits by delivering real-time, cost-saving measures that might involve rerouting transit vehicles around unexpected delays.

## Cities with Developing Public Transit Systems

Cities that are still in the planning or early development stages of implementing a public transit system stand to gain the most from quantum AI, for example, by using the technology to calculate the most cost-effective routes that could benefit the greatest number of people. Focusing on urban areas with higher population density along with linking those populations to other strategic areas of the city can benefit the city's entire population. Improved public transportation will increase citizens' range of movement within the city. This has pragmatic benefits, including greater ease of mobility for information access (especially in cities with limited access to research centers), increased access to more potential employers, greater access to educational opportunities, and even more convenient access to medical and recreational facilities. Without access to essential services, people may suffer from a lack of access to both healthcare as well as opportunities for personal and professional advancement. One of the benefits of implementing quantum AI in the development stage of a public transit system is that it may be able to calculate and organize a wider range of transportation options that would best serve the entire population.

In the construction of such a public transit system, quantum AI may be used in calculating the best ratio of expense to long-term benefit of varying combinations of transportation modes in the system. Regional weather concerns, expected population growth, and the distribution of essential resources throughout the city should also be variables in the model. Additionally, access to nonessential resources could also be considered in order to develop a system that ensures equitable access to the entire region. Interconnections to nearby cities should likewise be considered, especially in population-dense areas to ensure that maximum access to potential job and educational opportunities are available to all members of the community.

One transportation mode that could become an essential part of future public transit systems is the intercity hyperloop. A hyperloop could be especially effective for transport between cities in densely populated regions that would allow for enhanced high-speed mobility for people seeking work within their regional vicinity. This could greatly expand the scope of places to work and study for people who might need to stay in a region for personal reasons by providing a fast and efficient form of transportation, and quantum AI will be vital in carrying out the necessary analyses for such a complex and expensive undertaking as a hyperloop.

Quantum AI modeling may be used to more accurately estimate the costs per expected user, and this can be added into the analysis to determine what type(s) of public transit might be most efficient in terms of human, economic, and environmental impact, while also calculating optimal routes that will potentially benefit the most system users possible. Recognizing the importance of adjusting system analytics to reflect expected need based on community research is essential, rather than relying on demographic research alone; for example, considering what percentage of resident addresses in a particular area have driver's licenses rather than simply how many residents live in that area. Recognizing the needs of users in advance is essential for maximizing the benefits of a public transit system for the entire community, and quantum AI will be able to analyze much more data with many more variables and in a much shorter period of time than classical AI.

Additionally, this kind of community research should not be conducted only once, but must be continually re-evaluated to make sure that the data used for making ongoing adjustments to the public transit system is always aligned with community requirements. With quantum AI, these needs could be updated in real time to accommodate dynamic changes in community needs. Similarly, predictive modeling could anticipate potential upcoming modifications to the public transit system (such as rolling out new routes) that might best benefit users. By constantly adjusting, the needs of the community can continually be met in order to respond to the growth and changes that occur within the community.

With the threat of global warming, the value of modifying an existing system may be somewhat limited, but the potential for calculating alternative impact trajectories of variances within public transit systems could be essential in creating marginally positive impacts. Even small improvements like calculating minimal route alterations might be used to conserve the energy used by public transit systems' vehicles that could create positive change going forward.

Making public transit systems more efficient and affordable could also produce another positive trend and lead to the decline of people owning vehicles. If users relied on a mass transit system for day-to-day transportation needs, then an overall reduction of individual vehicle emissions would occur over time. The net positive environmental impact of this trend could be more significant than the corresponding increase in space needs in the public transit systems caused by a sufficient number of people foregoing their personal vehicles for more affordable options. Additionally, future public transit systems could accommodate unique transportation needs by incorporating self-driving, ride-sharing vehicles to allow for specific individual travel needs not served by regular mass transit routes. Quantum AI can balance the fares, needs, and routes of these individual users in order to maximize their efficiency while also accommodating the increasingly diverse needs of system users if more consumers abandon individual vehicle ownership for environmental or economic reasons.

One example of a future public transit system is being developed in Neom, Saudi Arabia. The Line is a planned futuristic city (with an estimated population of around 1 million) that will heavily integrate the use of artificial intelligence in almost every area of life. The Line is a new design for urban living space that is planned to be 170 kilometers long but will not include streets for cars. Moving between sections of the city will take place using an AI-powered transit system that operates on a layer immediately beneath the city's primary living space. Placing transit below the layer of the city where most people will live and work allows for walking to be prioritized. The environmental impact of this development will be overwhelmingly positive. Use of the public transit system primarily for longer trips will amplify the positive environmental impact of the integrated design of The Line.

Applying quantum AI to the public transit system in Neom will open up new possibilities that can be modeled around the world as innovations for increasing the speed and efficiency of public transit. This modeling need not be limited only to cities still in the planning stages but can include applications via adaptations of available public transit systems technologies, thus amplifying the potential importance of Neom's public transit system as a model for the future. Additionally, The Line's focus on environmental integration will enhance the public transit system's focus on minimizing and offsetting the overall impacts of the system on the natural environment, while simultaneously becoming an essential part of everyday life. Applying AI, and in the future quantum AI, in Neom will provide a model for other cities around the world on how to implement appropriate technology to improve daily life for people living in large cities.

## Ethical Concerns

While the benefits of quantum AI can be considerable, there are also ethical concerns that can arise with its varying applications, concerns that extend well beyond the literary dystopian visions of computer overlords. In a very real sense, quantum AI can be programmed to manipulate data to create resource bias. These biases may not accurately reflect the needs of the user base for the public transit system and must be made transparent. Additionally, system users should be protected from corporate data analytics that might, for example, factor in public transit use as a negative when making decisions on a consumer's access to credit resources.

Another issue could arise if quantum AI algorithms are not programmed with all the relevant factors applicable to the system's users. Low-income populations may not be provided with enough access points, thus limiting access to vital resources. Distribution of access points (i.e., subway stations and bus stops) needs to be spread across the city with more access points where demand for the transportation services may be higher. Anticipated needs may be best calculated by doing field research that would involve community meetings. Going directly to the communities that will use the public transit system will allow programmers to fully model the needs of the entire community. This is not a trivial task because a public transit system will only be the best that it can be when community input is used to maximize the potential benefits for the system's users—a point echoed by J. M. Taylor in "Should We Let the Machine Decide What Is Meaningful?" later in this volume.

Community meetings and surveys can be used to determine budget limitations, environmental concerns, the placement of access points, the frequency of routes, and the destinations available to provide equitable access for all members of a community. Once established, quantum AI can be programmed to respond to dynamic community feedback and provide optimizations based on different user communities as new information is collected in real time. These updates can then be used to adjust transit routes or available transportation modes (for example, buses versus ride-sharing vehicles). By doing this, quantum AI will be able to enhance the ability of a transit system to meet the unique needs of specific communities and enhance its overall usefulness for the entire community.

Quantum AI may be used to recognize and provide solutions for the efficient and ethical distribution of public transit system vehicles and routes that will create more equitable access to benefit the communities that the system serves. One pragmatic step that quantum AI could be used for is to calculate and create a fare system based on the unique needs of the communities served by the transit system. These fares can vary by route, point of origin, and even peak usage times, thus allowing for economically depressed areas to benefit from lower fares. Users who frequently use the public transit system could also benefit from recognition of their frequent use and receive fare discounts, which would especially benefit people using the public transit system to commute for work, as well as encourage broader system usage.

Quantum AI holds much promise for advancing society and lessening our impact on the environment. Both current and future public transit systems will be able to benefit from implementing quantum AI systems for optimization. While quantum AI can be used to address socioeconomic discrimination, it will be important to legislate restrictions on its use to ensure that those who use public transit systems do not encounter discrimination. The potential benefits of applying quantum AI to the construction and continuing development of public transit systems are profound, and this essay only begins to speculate on its wide-ranging benefits. At its core, however, is the belief that the best use of quantum AI will be to provide people with the most useful transportation options and simultaneously protect the natural environment.

CHAPTER

# 11

# The Road to a Better Future

Denise Ruffner, *Chief Business Officer, Atom Computing;* and André M. König, *CEO, Interference Advisors and Entanglement Capital*

Few people will dispute that we have firmly arrived in a world dictated by data. Facts are increasingly hard to verify, and yet certainty is the true north star. Opinion, bias, and memes drive not just behaviors but most buying decisions. This is as true for a protein shake as it is for enterprise technology. Hence, only those who do the following can succeed in today's world: 1. Collect data; 2. Clean and prep data; 3. Store, secure, and structure data; 4. Visualize and analyze data; and (this is the big one) 5. Act on it consistently and regularly. Welcome to the era of quantum AI.

The world is already divided into two camps—the data haves and the data have nots—and this divide will rapidly widen over the next several years. Note: you can't win at data with old database technology and some AI on top of it.

## The Future of Quantum Technology

The quantum technology ecosystem has grown tremendously over the past three years. Not only is the overall number of participants (vendors, startups, investors, etc.) rising at a fast rate, but the maturity and reach of each group has also grown significantly.

In the fall of 2020, there were close to 500 startup companies involved in quantum information science and its applications—with quantum computing hardware and software companies making up the bulk of that group. However, a strong base of quantum sensing, quantum communications, and quantum key distribution providers exist as well as a steadily growing number of service providers and media companies focused on the quantum information science space.

All of these technology development ventures are consistently moving closer toward commercialization, and the market for quantum technology is reaching new maturity levels. Yet, enterprises often struggle to understand quantum computing capabilities, preventing them from devising effective quantum computing implementation and deployment strategies.

The reality is that identifying business-relevant quantum computing use cases that can be realized in a reasonable period of time remains a challenge, as progress on the underlying quantum algorithms to deliver increased speed over classical approaches (the infamous exponential improvements that we all are clamoring for) remains slow. It is especially difficult in the context of enterprise computing with all of its potential challenges delivering innovation at scale and rolling out new technologies.

Current, embryonic quantum computing hardware—the NISQ-type machines—enable enterprises to trial the basics of quantum computing, allowing them to develop quantum competencies and validate early quantum computing applications. This is a crucial step on the quantum journey for any user, as, unlike other technologies, they will not simply be able to flip a switch or buy their way into it once they decide it becomes necessary.

The consequence of this situation is that despite all the excitement, rapid progress, and promises, the commercial market for quantum technology remains small today, and projections for the next decade demonstrate strong growth potential, albeit from a low base. According to the QIS Data Portal, revenue from all quantum information science vendors in 2020 is estimated to be just below $200 million USD with a 10-year projection of $2.2 billion USD in 2030. Slightly more optimistic growth assumptions put that number at $9.1 billion USD.

An emerging market is evolving where the technology is developing but still not showing prominence over competing technologies. Enterprises are struggling to understand quantum computing capabilities and business-relevant use cases. However, there is little doubt that the technology will have a significant economic impact once realized.

## Commercial Near-Term Impact

The economic impact in terms of revenue potential is relatively small in the immediate future, so as large vendors and original equipment manufacturers (OEMs) pour resources into quantum technology and investors evaluate quantum information science ventures, this means the return on investment will likely come from somewhere else.

Private investments in quantum information science ventures have been steadily rising across the range of quantum information science applications, yet compared to other technologies, the private investment total remains relatively low with a total projected $1.5 billion USD in 2020 according to QIS Data Portal.

These dynamics will lead to significant consolidation among vendors and solution platforms as more and more teams will struggle to justify the viability of their efforts. This is natural in an emerging market and a sign of growing maturity. Current vendors are focused on selling cloud access to quantum computers, machine time, software, and consulting services. There is a natural revenue limit to this approach in a deep technology environment, especially since quantum computing companies need to realize that they are competing not just with each other but with other classical enterprise technologies.

Quantum AI opens up a whole new realm of possibilities. Imagine a delivery company able to optimize their delivery routes and save 5 to 10 percent in gas costs, fleet maintenance, and employee time. Or imagine a pharmaceutical company able to predict which ligands in a compound library would best fit with a receptor of interest, thereby greatly decreasing the time to develop a new drug. Or imagine a bank able to improve its fraud detection through quantum machine learning algorithms thus decreasing revenue loss. While these all are valid use cases, it remains to be seen if breakthrough results can be delivered within a reasonable timeframe, and not just because hardware, software, and algorithms need to improve significantly to facilitate this objective. For example, a large manufacturer might not choose to switch its entire production lines and processes for a 10 percent productivity increase, especially when much of that improvement can be delivered using classical methods over time, with less risk, and with more predictable budgets.

Enterprise adoption, thus, is not a theoretical concept rooted in quantum supremacy, but rather an exercise in weighing the pros and cons while carefully calculating the return on investment. Usually, a 10 to 20 percent improvement in a process, product, or solution may not be sufficient to drive adoption, so, therefore, unless quantum computing is able to deliver a significantly larger change in productivity or innovation, most enterprises will be slow in implementing quantum solutions, thus limiting the revenue potential for many current business models in the quantum technology ecosystem.

So what will provide growth to sustain the quantum ecosystem for the long term?

## Finding New Sources of Revenue

Economic impact in terms of new revenue sources becomes more significant as users adopt quantum computing with increasing vision. Success here will hinge on hardware devices growing in quantum volume, and related artificial intelligence software expertise being developed to address novel use cases. This could lead to more than just cost or performance advantages and produce new revenue growth. When looking at the types of quantum algorithms (i.e., prime factoring, Monte Carlo, optimization, variational quantum eigensolvers, etc.) that might deliver a speedup over classical computing and mapping them to their potential industrial use cases, clear solutions around Big Data and machine learning applications emerge.

This implies a switch toward "solution selling" rather than selling proof of concepts or consulting, as well as a better understanding of business models, enterprise sales, and corporate strategies. As vendors position themselves to really address some of the large problems and offer comprehensive solutions to them that go beyond "just quantum," the industry will see much larger commercial projects' engagements.

The impact of these projects has the potential to deliver more than just efficiencies and, in fact, enable new business models or competitive advantages for the user. There will be disruption among corporations that have invested early to develop quantum computing solutions that will have an advantage over late adopters. The significance of this will be realized and should be part of every company's planning as they develop their own quantum strategy.

## Disruptive Innovation—Where Is Our Hero?

Economic impact in terms of entirely new revenue streams is the big unknown. It will likely take an "Elon Musk of quantum computing" to come along and upend the entire industry. Entrepreneurs will need to have the courage and fortitude to dream about science fiction types of innovations (cold fusion, energy independence, radically new materials, financial innovations) that might potentially create trillion-dollar business models.

As large OEM companies will certainly be successful in driving cloud and services revenue, breakthroughs—the quantum revolution—will ultimately come from aggressive and well-funded startup companies free of business model constraints and ingrained processes and behaviors.

But for that to happen, first a few things need to occur.

Entrepreneurs need to dream big—we know few founders in quantum computing that are truly thinking big.

Not only do these entrepreneurs need to think big; they need to effectively communicate and position themselves through roadmaps and deliberate steps toward achieving their goals. Unless you are sitting on your own pile of cash, this type of disruption does not take place in a dark lab three levels underground (or is there a government agency already working on it?). In essence, you need a compelling story to sell.

Investors will need to adjust their appetite for increased risk. Venture capitalists have become so professional and well-honed that there is little room to peek beyond their spreadsheets. As long as investors are required to assess a venture based on traditional key performance indicators, the funding for such a moonshot company will be hard to come by.

It is also the aforementioned chutzpah of the yet-to-materialize founder that will enable them to hire the highly motivated and dedicated team of scientists and professionals necessary to realize the company's vision. As of today, the best entrepreneurs are the most methodical, taking the intellectual property out of their research labs and cleverly commercializing it. We by no means condemn this approach to innovation, as, in fact, it is the driving force behind much of the current growth within the quantum technology sector. Nonetheless, we do hope for a pioneering type of leader to emerge and have a go at it—ushering in a new industrial revolution.

Imagine helping the world solve the carbon capture problem and how that might change our future. Imagine curing diseases, providing unlimited and safe energy and developing real general AI. We must also imagine terrible weapons of mass destruction. These are all possible. It takes a dream, a plan, funding, and technology to get there—with a strong, global regulatory and oversight framework to avoid the potentially negative effects.

We also want to emphasize that the new future workforce, as well as the current workforce, needs to gain an understanding of quantum computing. It could be from a programming point of view or from a corporate strategy point of view. Nonetheless, people conversant with this technology are needed, and this need will grow only as the technology grows. Educating our young people in this technology is essential as it is critical to our future.

The development of quantum computing will become a competitive edge for organizations and nations. We can see how much funding is pouring into this technology at national levels—Germany alone, in combination with the EU, has committed close to €4 billion to quantum information science. The economy of a country can vastly change with innovative quantum solutions. However, we also need to consider that as this technology gains in capabilities there is a possibility that governments will limit sharing of the technology, as it could impart defense and military advantages.

## The Future of Quantum AI

Quantum computing is the only technology coming into existence today with the potential to adequately deal with all of our data and the challenges that we are facing. This has, to a certain extent, to do with the raw power of these machines, and the more powerful they get, the more they will be able to solve. But, and this is the crux, it also has to do with the nonclassical nature of our world. Many problems, when we try to describe them as humans using math formulas and equations, aren't rooted in classical physics. They are rooted in quantum physics, and you cannot fit a square peg in a round hole. The message here is simple: the world is changing. We must accept this change. To strive in the new world, mastering data is the only answer. Mastering data requires computational capabilities that go well beyond classical high-performance computing, blockchain, or AI—not in speed, but in nature. It is time for everyone to start practicing that quantum nature and become quantum intelligent.

PART

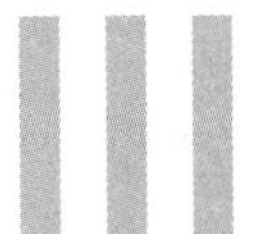

# Social Impacts

CHAPTER

# 12

# The Best Numbers Are in Sight. But Understanding?

Roald Hoffmann, *Frank H. T. Rhodes Professor of Humane Letters, Emeritus, Cornell University;* and Jean-Paul Malrieu, *Director of Research, CNRS IRSAMC, Université Paul Sabatier, Toulouse*

The chapter title voices our skepticism, and given that we are both in our 80s, it is easy to attribute it to our age and attendant creeping conservatism. But in a time of hyped enthusiasm for the New Jerusalem of IT, we thought there might be a place to question the systematic confidence voiced and eventually define a more balanced, if oblique, perspective.[1, 2, 3]

Our response to what so-called artificial intelligence has done and will do to our lives is complicated. We want to rehearse with you the pretty obvious reasons why this is so and then go on to what really worries us. This is the attack—scientific, philosophical, and psychological—that artificial intelligence, augmented by quantum computing, might represent on a human jewel, the idea and processes of understanding.

## Attitudes Toward Quantum Computing

It's hard (and would be strange) to be against quantum computing. Quantum computing is marvelous on several accounts. To those of us who devoted their lives to solving approximately the wave equation of quantum mechanics and connecting it to the very tangible world of chemistry, it's astounding to see superposition and entanglement turn into operational reality and precise numbers. We thought of those inherent quantum mechanical notions more as philosophical quandaries, rather than the domain of human quantum mechanics such as ourselves. And here, today and not tomorrow, these concepts are put to practical use. The design of the physical building blocks of a quantum computer—the qubits—also returns us to chemistry. Realizable qubits may be inorganic molecules, or an addressable defect in a crystal. One worries about state fidelity and decoherence, and so a new bridge between chemistry and computing forms.

## . . . And Artificial Intelligence

Decades of life with computers have cleared the way for artificial intelligence to enter our lives. To search a library's holding for the title of a book it contains, to plot out a wave function—those are computing tasks, easily described for a human working with more primitive tools, easy to understand how the computer does it, since we wrote the algorithm. And it is always marvelous to see the computer's speed and to smile at how easily our mistakes mislead the machine.

On to machine translation from a foreign language, and our reaction is still unsullied wonder at how well Google Translate and deepL can do it, using a neural network framework.

We go on to facial recognition, and now things become murky, not because of lack of admiration of how well the program does it, but because of the use to which human beings and machines may put these programs. Let us be specific: Cliff Kuang has described clearly a study by Michal Kosinski, using data from dating profiles, correlating sexual preference in individuals with self-identification, based on images of their faces and self-supplied profiles.[4] The data scientists achieved a much better correlation than human evaluators. Now while that may seem harmless enough (but could some institutions evaluate the suitability of a person for a job based on that "identification"?), the stretch to misuse in the treatment of certain groups, say the Uighurs in western China, as a prelude to "re-education," is not a long one.[5]

So, the problem is not in the machine, but in human abuse. Two comments on this:

- So, human beings are fallible. Given any technological innovation, a certain fraction of the human users is bound to misuse it—for their own gain, to shame or harm others, or just for irresponsible creation of chaos. Surely the best way to counter this is to have the legal structures and strictures to limit and guide behavior on the computer.

  Of course, we must try. But realistically, human misuse (whether of chemistry in synthesizing to order new, addictive opioids, or plagiarism) always outpaces in its ingenuity and scope the rules and regulations we impose. The mental, not physical, energy that powers hackers and trolls is hard to match in a defensive stance.
- Maybe humans should get out of this and counter automated computerized misuse by computerized protection. No, the

ethical situation is not resolved by leaving the choice in the hands of programs or machines. Remarkably, this argument has been made in another context:

> "Several military experts and roboticists have argued that autonomous weapons systems should not only be regarded as morally acceptable but also that they would in fact be ethically preferable to human fighters. For example, roboticist Ronald C. Arkin believes autonomous robots in the future will be able to act more "humanely" on the battlefield for a number of reasons, including that they do not need to be programmed with a self-preservation instinct, potentially eliminating the need for a "shoot-first, ask questions later" attitude. The judgments of autonomous weapons systems will not be clouded by emotions such as fear or hysteria, and the systems will be able to process much more incoming sensory information than humans without discarding or distorting it to fit preconceived notions."[6]

This argument is weak: the robots may be programmed to survive, since they are expensive and may be employed in the next task that requires no judgment.

Taking a life, even just hurting someone, cannot be delegated to an algorithm. With the action comes moral responsibility, our deepest one. Were we to pass to that world—of machines making decisions of life or death—then our society has failed us, and we it.

## The Wave

Actually, what bothers us most about the wave of machine learning and neural networks, of artificial intelligence, that is breaking over our heads, is that AI makes epistemological claims.

The claim is that artificial intelligence provides real understanding. One needs to get the numbers (the energy of a molecule) or the face right in a recognition task to say that one really understands. Even if this were true (see the next section), does the correctness of the answer provide an explanation? As René Thom said, "Prédire n'est pas expliquer/To predict is not to explain."[7]

Let us furnish the background for this assertion by telling you what we mean by understanding and place us in a line of theory that has understanding as a goal. Then the attack of AI on understanding will be clarified. We will also outline a psychological aspect of the struggle.

## Understanding

Elsewhere, we have gone at some length into the attributes of understanding. It is often tacit, a state of mind. And most of the time it's qualitative, though it can certainly have a quantitative aspect to it (and that is where quantum computing comes in). In physics and chemistry, understanding usually resides in theory and in interpretative models, a practice of analysis of anything observable in terms of the possible physical or chemical mechanisms (causes, elementary actions) that could lead to the observable and an estimate of the role of their contributions.

There is a strong pedagogic aspect to understanding. In fact, one of the defining aspects of understanding is that it can be taught to an intelligent student (forget the professor; often they are unteachable), in words or concepts, in equations. This quality, the teachability of understanding, is what we use all the time to keep honest our colleagues who are aficionados of machine learning or neural networks. We probe, "What have you learned from your calculation that you can teach me?"

## Where We Came From

Something must be responsible for the skepticism toward AI from two theoretical chemists, whose whole professional lives were enabled by the progress of electronic computers.

We come from quantum chemistry. About 90 years ago, a sage of our tribe, P. A. M. Dirac, wrote this:

> "The underlying physical laws necessary for the mathematical theory of a large part of physics and the whole of chemistry are thus completely known, and the difficulty is only that the exact application of these laws leads to equations much too complicated to be soluble. It therefore becomes desirable that approximate practical methods of applying quantum mechanics should be developed, which can lead to an explanation of the main features of complex atomic systems without too much computation."[8]

We have been trying. But far from brute-force applied mathematics, we have tried to construct frameworks that allow us privileged passages to forming explanations. And, in another direction, we have tried to connect to qualitative ideas that chemists have formed, about the ability of atoms to share electrons, and of the varying propensity of nuclei to hold on to them and to link these to insights from quantum mechanics. Here especially useful was perturbation theory, a time-honored way of getting physical insight (understanding!) of equations that could be solved only approximately.

The sound of the old key punch is in our ears. As is Fortran. We became pretty good at what we do, meaning that we could calculate some numbers approximately, interpret (understand!) the numbers as factors influencing a real outcome, and, ultimately, construct explanations. Chemists, ever so talented at making molecules that effected a small change of atomic composition (an H here replaced by an F, and an $NO_2$ group by a CN), could then test our qualitative prediction. So ensued the most wonderful aspect of science, the dance back and forth between theory and experiment.

When it worked (and it didn't some of the time), people understood.

## Does the Machine Understand?

Machine learning and neural networks, new engines of simulation, arrive. We, who solve Dirac's equations approximately (but do much more), are deemed to have been replaced. Not entirely, for we are needed to get exactly right the energies of a goodly set of molecules involving the same elements. This is the "training set." In machine learning, a theoretician might also design the indicators that the AI machine uses to make its correlations—these will show the identity of the atoms, their positions in space, and other characteristics. And then the program, whether it is in supervised or unsupervised machine learning, is set loose to find its own way to the best fit of, say, the energy of the training set molecules.

The outcome is a predicted energy for an unknown molecule, one outside the training set, that is lower (better) than our present approximate quantum mechanical approach can give us.

You can see what the journalist writing the press release on that work will say: "AI now understands the molecule better than any calculation." The scientist often does not make the claim as patently, but when you mix the justifiably enthusiastic scientist with their institution's journalist and the news-worthiness-hungry editors of *Nature* or *Science*, you get the perfect storm for (to put it mildly) exaggerated epistemological claims.

### Psychological Factors

Roald thinks another danger lurks in the practice of simulation, which creates a block to even imagining that explanations, in the time-honored way of theory, might even exist. And that barrier derives from the psychology of human-machine interactions. Computer programs are naturally complex, made up sometimes of thousands of lines of code. There are always problems in getting them to work; the chore of debugging is an experience many of us have shared. If all that work has to be done to get a number, surely there cannot be a simpler way to get it, even approximately?

There is no question that computers are so much more efficient than human beings in pattern recognition. And it could be that human beings fall for simplistic explanations to avoid complexity. In their lives—and in nature—sometimes we think that the current wide range of belief in conspiracy theories, no matter how cockeyed they might seem, comes from that desire for a simple, ordered world. Nevertheless, we think the process of getting computer programs to work predisposes the programmers to depart from the hard path of finding a theoretical explanation.

### What Will Quantum Computing Do for Understanding?

For small molecules, quantum computing has solved those governing equations more accurately than any other calculation. It will soon do it for larger molecules. The best numbers will be provided by quantum computing, but true understanding, in our opinion, will not.

## Let Us Claw Our Way Back

Are we done maligning and moaning? First, the problem of AI providing numbers but not explanations was recognized early on. Part of the field is moving on to crafting the programs to tell us how an AI implementation (machine learning or neural network) learns and how it does what it does. This is an outstanding problem, that of "explainable AI." The U.S. Defense Advanced Research Projects Agency even has a program in the field. One does want to know what happens inside the box—to know what the computer does at its structural level and perhaps to learn from it something in the way of discernment or proof strategies.

Coming up with regularities in numbers is easy. Defining the regularity in terms of a classical theorem (say one of Ramanujan's incredible series) is a challenge. Finding a theorem that is "interesting" to a human mathematician is harder still. But there is progress even on this.

Explainable quantum computing? The beauties and constraints of error correction, the effects of decoherence, can be explained. But the result, a number, seems to be inherently mute until a human being (or maybe an explainable AI program) builds a story out of several or many such numbers.

## Seeking Numbers, Forming Theories, Creating Narratives

Like it or not, the future will hold much more simulation and AI than the two of us would like. At the same time, both of us have confidence in our students—they will find a *modus vivendi* in theoretical chemistry in that future world. We'd like to think about what that world would be like, with at least a partially open mind, and perhaps find something special in that future.

Our own experience with numerical calculation (call it simple simulation) and the building of theories gives us a clue. If we look at it abstractly, we see (for both us, although we are quite different scientists because of our education and history) a similar dynamic. We alternated spells of highly detailed quantum chemical calculations with the construction of simplified explanations that involved our game pieces, orbitals. These orbitals, or rather the electrons in them, are involved in very specific interactions. These we puzzled out, sometimes directly, sometimes with more than 20 years of work, through specifically tailored probing calculations—simulations with an aim to learn something from them, not necessarily to simulate reality.

If you allow us the conceit of thinking of our numerical calculations as simulations and grace them with the AI label (how we would have blanched at the thought!), then we alternated periods of AI simulation with theory building. Lo and behold, it worked! At the end of it was a double satisfaction—a qualitative theory to explain the chemistry, and a prescription for the level of theory needed to get specific experimental numbers right.

We built a story out of numbers and theory. But our students will do this better, dazzling us with how they jump in and out of riffs of computation.

Stories are ancient; stories are deeply embedded in the human psyche. When the wooly mammoth was killed, the hunters told in reliable geographic detail where the beast was found (cross the river, turn right at the giant oak, go up beside the cliff). And, we suspect without a pause they recounted the fierceness of the cornered giant, the courage of the hunters it trampled.

Theories are stories. They share with fictional tales temporality—a calm beginning, a problem to be solved in a tense middle (where you don't know if the applied technique will work, so stumble toward another), and an ending, which itself carries an inherent tension. One does want to give the impression that something significant was learned and yet must leave the reader/listener with a feeling that there yet remain mysteries still to be solved. Theories clearly ascribe causation—they are the most deterministic of narratives.* Thinking one has seen all the causes (an almost excessive rationality) is actually a problem for science.

And do scientific stories have human interest? Oh, they surely do. Sometimes we need to be privy to the cognitive structures of the field to appreciate them—as in Einstein's remarkable use of the entropy of radiation in his classic 1905 photoelectric effect paper. Sometimes we need the rivalry of competitive theories, with real people pushing them on. Sometimes a theory is so compelling that it needs little new experimental support—we think Darwin's theory of evolution, perhaps the greatest story ever told, was like that.

## Looking to the Future

We think that AI and quantum computing will enter the chemistry of the future in two ways. The first is in their simple utility—in perfecting the technological capabilities of the chemist to make any specific molecule or material or to design and make a molecule with certain desired properties. The aim should be to serve humanity, of course, even as we are aware that abuse of that wonderful synthetic capability of human beings is, sadly, common.

But our spirit wants more. We want to *understand*, and yet quantum computations will remain mute. But explainable AI has a future—a theoretician will use it to find regularities worth thinking about and will explore their origins. They will come up with better frameworks for understanding, frameworks that can then be taught.

---

*For more on storytelling in science, see Hoffmann, R. (2000) "Narrative," *American Scientist*, 88(4), 310-313; Hoffmann, R. (2005) "Storied Theory," *American Scientist* 93(4), 308-312; and Hoffmann, R. (2014) "The Tensions of Scientific Storytelling," *American Scientist*, 102, 250-253.

## World Building

We imagine a world in chemistry and other sciences as moving toward near-infinite capacity in the *technê* of searching for facts and properties. And also a world where we hope our students, and not just a privileged few, will experience, working hard all the way, the *sophia* of making sense of things.

And now a flight of fancy, an excursion into the imaginative realm, appropriate to the origins of this book and our debt to Jules Verne and Ursula LeGuin.

Elsewhere we have imagined a Museum of Science of the future, no doubt virtual, as the current pandemic has gotten us used to. We describe the objects in it:

> "In one room hangs the Cassini mission image of the lakes of ethane on the surface of Titan. In another, the discovery of archaea. And of how their lipids differ from ours. In a third room we see Onsager's solution of the two-dimensional Ising model, in a fourth room the synthesis of coenzyme-$B_{12}$."[13]

This is a sacred space, these are artistic and scientific achievements, of pervasive *spiritual* value.

Does that seem a long jump, from coping with AI to the spiritual? Or just two scientists going soft? Let us trace the chain of thinking that moved us there.

Our lives and our science have been transformed by information technologies, with more change on the way. And there are dangers around us, from real anthropogenic climate change to totalitarian tendencies masquerading as populism. The two of us differ in our degree of optimism about the human condition. Will we in fact be sage, build safeguards, and form more just societies? Or will too many perish along the way, leaving only a scattered remnant; humanity may survive, but at what cost?

Whatever happens, artificial intelligence will be an essential part of it. And . . . there will remain human beings who long to experience and be motivated by more than ultimate efficiency and accurate numbers and who will seek true understanding. The space of understanding in us is the same space that is touched by music, poetry, and all the arts. It is a sacred, deeply human space.

The theoreticians of the future will be listened to because they will have learned how to tell their stories of theoretical discovery in a convincing narrative. It is not out of place that some will take Carl Sagan's pointer and use fiction. And—ideally not just wishful thinking—some of these theoreticians will also be master teachers, recognizing the intertwined ways of understanding and teaching.

The aesthetic aspects of chemistry will always be there—the stinks, bangs, and vivid color changes that attracted Primo Levi [9] and Oliver Sacks[10] will be brought up-to-date by the heirs of Theodore Gray [11,12] and Yan Liang.[13] The sheer variety of properties and function in the denumerable yet astronomically (no, chemically) large set of possible molecules will pull in people, as will symmetry and the lack of it, the power of frontier orbital reasoning, and the sharp logic of organic synthesis.

Just thinking about all that beauty in variety sends shivers through us. This is perhaps the evidence that the logical has caught sight of the spiritual. The husk covering the sublime has been breached.

We see the future theorist making use of AI and quantum computing to play infinite games. The games' purpose is not their reliable numbers, but the stories that the theorist can assemble from all those explorations of the "what if?" world. The best stories that emerge from this directed roaming will please you and touch you. Perhaps then the distinction between art and science—those different ways of knowing this world, of *understanding*—will become less important. Because both touch the spiritual in us.

CHAPTER

# 13

# The Advancement of Intelligence or the End of It?

Kate Jeffery, *Professor, School of Psychology & Neuroscience, University of Glasgow*

Of the universe's roughly 13 billion years, the last four billion years or so have seen an extraordinary development: the evolution of life on Earth. First came simple unicellular life, then simple multicellular life, and finally complex multicellular life, culminating in nervous systems and eventually cognition. The pinnacle of this process (so we like to think) is human intelligence, which has allowed us to create science and technology and so vastly expand the scope and scale of life's operations. We have even managed, probably for the first time in the history of terrestrial life, to escape the surface of the planet. Now, we humans have used our ingenuity to create a new type of intelligence that depends not on biological molecules but instead on silicon-based computing and is tapping into new domains of physics that even life has not yet discovered. Although we have not yet cracked the problem of creating a truly intelligent artificial intelligence (AI), it is surely only a matter of time. Technology is advancing rapidly, silicon-based evolution is surpassing carbon-based evolution, and the future possibilities seem boundless.

Then what? Are we on the brink of a glorious new era of evolution in which intelligent life in the universe is no longer carbon-based? Will our silicon-based descendants venture out into the cosmos and become interstellar wanderers, boldly going where no life has gone before, conquering new planets, and seeding intelligence across the galaxy?

This is a pleasant prospect, but let's examine its likelihood. Whatever our fate, there is one inescapable fact, which is that it will be governed—as everything is—by the natural laws of the universe.

## The Statistical Mechanics of Life

The fundamental problem that disturbs our pleasant dream of galactic colonization is the second law of thermodynamics: entropy, which governs the unfolding of the universe over time and cannot be evaded. Entropy is often taken to mean the process whereby systems become more disorderly and less structured over time: eggs break, the froth on coffee disperses, ice sculptures melt, gin evaporates. However, entropy actually isn't about order—it is about the (rather circular) tendency of more probable things to become more prevalent over time. It just so happens that usually disorderly things are more probable than orderly things. Given an egg and all the states it can be in, there are many broken states but only one formed state, and so eggs break more readily than they form.

The egg *did* form at one point though—inside a hen—and this is because in certain situations, given the properties of matter and a particular starting point, an orderly state is actually more likely than not. When oil is mixed with water and then left to its own devices, the disorderly mixture evolves into an orderly separated state with oil and water in neat, separable layers. Because of gravity, an orderly state of separated oil and water is more likely than a disorderly mixture. Life is a spectacular example of order forming from disorder. Given a healthy mature hen plus the laws of chemistry and physics, eggs become highly probable (sometimes too much so, as any chicken-keeper knows). All these processes are nevertheless entropic. Combined with natural selection processes that allow the most resilient of these newly developed orderly structures to persist, complexity in the universe slowly grows over time. There is reason to think that we humans and our machines may be the most complex system that has yet evolved.

Entropy and energy are tightly intertwined. When systems change state, moving from the less-probable state they started in to the more probable one they end up in, energy is used up—which is to say, it is converted from a form in which it can do work (move things), which we call *free energy*, into a form in which it can do less or no work. Free energy is able to do work because it is at the top of an energy gradient—a ball at the top of a hill, hot coal next to cold water, etc.—and where there are gradients, atoms can be moved, and work can be done. After the work has been done, the gradient exists no more, and that's the end of it. This running down of free energy is what we mean by entropy.

As mentioned earlier, a consequence of the tendency for order to arise spontaneously from chaos, driven by the entropy probability engine, is that certain things in the universe become more complex over time, which is how we ourselves were able to come into being. However, when the universe generates more complex entities, as it slowly evolves, these entities consume more and more free energy due to their more complex behaviors. Like all entropic processes, this energy is turned into heat, which dissipates and becomes unable to do further work. When atoms bond and form a molecule, heat is released. When matter condenses to form a star and atomic nuclei start fusing, vast amounts of energy are turned into heat. When matter forms self-replicating molecules that discover how to photosynthesize, they produce heat. The story of life can be thought of as the story of the universe discovering, driven by entropy, multifarious new ways to turn free energy into heat. By enhancing the development of complexity, entropy thus speeds itself up.

## Information

Entropy and energy also interlink with a third quantity, information, which is where intelligence comes in. We can think of information for our current purposes as connections to the energy gradients in the surroundings. To the extent that one knows there is energy *here* and not *there*, one has information that can be exploited to use that energy gradient to enhance survival. Life's first use of information was to figure out how to order the arrangement of nucleic acids so as to harness chemical energy to sustain itself. This was a self-enhancing process that steadily accelerated, shaped by the process of natural selection, which is another variant of the law of entropy in which more probable—or "successful"—life forms become more prevalent over time. As life evolved, it began to process many different kinds of information, and as it did so it acquired organs to enable this: first intracellular organelles, but then—when cells began to aggregate and cooperate—specialized tissues, and then neurons, nervous systems, and finally complex cognition. Each of these developments enhanced the ability of the organism to exploit energy gradients, speeding up the conversion of free energy to heat.

Information processing has reached a pinnacle in humans, whose evolution has been characterized by the sudden acquisition of a new way of gaining and exchanging information: language. We don't know when or why human language evolved—this capacity seems to be unique among living things—but when it did, it enabled a whole new layer of information processing to be added to the action repertoire of these otherwise unremarkable primates. The power of language is that because it can be transmitted from person to person, the information it conveys can persist in the population and build upon itself—ideas no longer die when the brain that holds them dies. This was a momentous step forward, and it has propelled us from our hunter-gatherer origins to being the urban, planet-enveloping space-exploring technophiles that we are now.

One of the things enabled by language was the development of new ways of processing information that go beyond our own brains (although brains are, themselves, impressive information processors). A notable step along this pathway was the development of writing, which meant that there was no longer a need for direct contact between people for ideas to be transmitted between them—ideas could now spread far in space and persist far across time. The invention of the printing press vastly accelerated this process and brought previously unimaginable amounts of information directly into the homes of ordinary people. Another major transition has been the one that we are living through at present—the invention of personal computers, which mean that almost all the information that has been collated by humans, past and present, is available to any individual with a computer and an Internet connection.

The pace of information-processing technology continues to increase. Riding on the wave of the computer revolution, and currently hurtling toward us at great speed, is artificial intelligence. This means that not only can we transmit enormous amounts of information between ourselves, allowing billions of brains to operate on that information and perhaps use it to advance technology, but we have the potential—if the expectations of AI come to fruition—to develop new brain-like processors that can share information directly between themselves and cut us out of the loop.

## Two Problems with Information —Energy and Viruses

This is all very exciting. However, information, and the ways in which complex life-forms have found ways to use it, has two big problems that threaten to limit the capacity of our species to advance our technology and venture out across the galaxy in the way that we have fantasized. One is its connection to the entropy engine; the other is the tendency for information systems to prey on each other.

The issue with entropy is fairly straightforward to understand—information links to energy, and energy to entropy. Information processing is one of those behaviors of complex systems that speed up entropy. Information both consumes increasingly large amounts of free energy to process, and enhances our ability to use energy for furthering our growth, in the manner of all life forms. For example, it has been estimated that the Internet accounts for as much as 2 percent of civilization's current energy consumption, and this is growing exponentially. In its vast capacity to enhance the information processing of our species, the Internet enables new and clever ways of turning free energy into heat. A spectacularly self-harming example is cryptocurrency, whose sole function is to turn energy into information and in doing so to influence that other vast information-processing device of humans that we call the economy. This process is accelerating, and the energy consumption of cryptocurrency is skyrocketing year on year, currently consuming as much power as the whole of Estonia. On Earth, because of the way living things work and our current over-reliance on carbon-based fuel sources, the turning of energy into information generates not just heat but also carbon dioxide. This blankets Earth and prevents its excess heat from escaping into space, thus warming the planet like a huge duvet. Fueled (literally) by this and other carbon-producing activities, the current pace of global warming is accelerating, with no sign of even a slowdown let alone a reversal.

The second problem with information (as if the first were not enough) is that when information systems develop, they very soon start exploiting each other. The most primordial version of this arose when the first nucleic acids began to self-replicate. Almost immediately the first viruses—strands of nucleic acids themselves—appeared and began to interact with them to harness their energy-producing systems for their own replication. Viruses still plague us today—tiny packages of DNA or RNA, wrapped in protein, whose sole function is to infiltrate the cells of other organisms and coopt their replication machinery to make more of themselves. They exist solely to exist—which is true of all life, to be fair, but viruses really lay bare the fundamental circularity of existence. This is, of course, entropy—chemical systems turning free energy into heat. Their self-replicating nature is a reflection of the process we discussed earlier: that entropy drives complexity, and complexity drives entropy.

Viruses aren't just made of nucleic acids—any information system that colonizes another system and hijacks its replication machinery to make more copies of itself could be called a virus. Very soon after the widespread introduction of personal computers, when computing information systems started to interact, the first computer viruses appeared. They are a constant hazard, and detecting and thwarting them is a vast information-processing enterprise in and of itself. Antiviral systems are sophisticated, but viruses are constantly evolving, and it's only a matter of time until a silicon-based virus brings our global economy to its knees, in the same way that the carbon-based viral pandemic that is raging at the time of this writing.

We also use the term *viral* to refer to human ideas—cognitive constructs that spread, via the replication machinery of language, from one brain to the next and propagate themselves. Richard Dawkins gave the word *meme* to these gene-like cognitive fragments that spread from person to person—that word itself has become a meme. If you hadn't encountered it before (which seems unlikely), then you have now been infected with it—and if you find the idea interesting, you may spread it to someone else. As noted earlier, the spread of ideas is what has driven our civilization. However, the replication machinery of human ideas—language—opens up a portal to colonization by virally spreading self-replicating ideas. This colonization can be harmless—often beneficial—but not always. The vaccine hesitancy meme is currently spreading like wildfire across the world and threatens our ability to fight off traditional, carbon-based viruses like measles, polio, and of course SARS-CoV-2. Since the development of the

Internet—another replication device for ideas—we have been infested by a plague of "fake news." Fake news consists of false ideas that use the Internet to spread across the globe, seeding human minds with incorrect or downright harmful notions that threaten to undermine our societal stability and (relative) harmony.

Viruses are an inevitable side effect of replicating and interacting information systems, but they constantly threaten to destabilize those very systems. They are a product of entropy and a reflection of how the complexity that is driven by entropy is always teetering on the edge of collapse.

## Quantum AI and Viral Entropy

This all brings us to quantum AI—the much-hyped technological breakthrough that we are poised on the brink of. AI—artificial intelligence—is the name given to computer systems that set out to mimic human intelligence so that we can turn over some of our tasks to machines and perhaps even have them solve problems that are beyond our own capability. Traditional AI has been hampered by the slowness of classical computers relative to the vast number of processing operations required to simulate even simple brain functions. The brain is massively parallel, but a traditional computer has to do everything sequentially, which greatly slows it down. Adding to the confusion, we also don't yet have a comprehensive understanding of how brains work, which makes mimicking them even more challenging. Nevertheless, AI has been making impressive strides in recent years, fueled by the accelerating pace of microchip technology and processor speed.

Quantum AI works—in theory, although not yet in practice—by capitalizing on the ability of quantum systems to exist in multiple states simultaneously, meaning that a quantum computer can parallelize computations and so relatively quickly perform some types of calculation, such as breaking an encryption code, that would take a conventional computer a prohibitive amount of time. Quantum technology is in its infancy and requires vast amounts of energy to sustain the ultra-low-temperature processors used to maintain particles in a state of quantum entanglement. Currently, therefore, quantum computing is like nuclear fusion inasmuch as the promise of unlimited power, be it electricity or computing power, is still a theoretical possibility rather than a technically achievable reality. However, first steps have been taken along the road to practical implementation of both of these technologies, and in 2019 Google announced that it had achieved "quantum supremacy," in which a quantum computer solved a problem too complex for a classical computer (although there is still debate as to what exactly constitutes quantum supremacy and whether Google did indeed achieve it). It is time, therefore, to consider what the future would look like with quantum AI at our fingertips.

You are probably starting to get a flavor of where the foregoing arguments are leading us. Each one of humanity's technological developments has had one ultimate aim—to increase replication of the system that created them. These developments are just complicated ways of turning free energy into heat by manipulating energy or information or both, and like all information-processing systems, they are energy-consuming, unstable, and prone to viral infection. We can dream about AIs that can do the dishes or walk the dog, but the reality is that as soon as AIs are developed, they will be weaponized, as with all other technological developments, and used by us to kill each other. They will also be enormously costly in energy, which means they will accelerate the pace of global warming. The end result will be that civilization dissipates as heat, one way or another.

Are we, then, thermodynamically doomed? After all, complex systems are just entropy's way of making faster entropy, and quantum AI will be the most complex system yet. Statistical mechanics says that we must eventually return to the formless cosmological soup from which we sprang, and we are constantly accelerating our rush toward this state. The path from here to there, though, is not fixed, and perhaps it can be shaped so as to buy us some time.

## Solutions

The problem of runaway weaponized information systems is not a new one. As mentioned earlier, almost the moment the first speck of life appeared on Earth, the first virus arrived to exploit it. However, the story didn't stop there. Not long after the first virus appeared, the first antiviral system appeared—DNA repair mechanisms that look for disruptions to the genetic code and sort them out. Then viruses got cleverer, but then so too did cells. As living systems developed in complexity, so did their self-preservation mechanisms—not only do we now have an intricate machinery devoted to keeping our genetic code intact, but we also have an extremely complex and effective immune system that patrols the body and detects and destroys invaders. More recently we have evolved molecular biologists who add another layer of protection, via vaccines, to our armamentarium. Even so, viruses, and that other viral cellular problem, cancer, remain our biggest problems, because viruses are always changing and adapting. We have seen the same arms race with human technological developments—as computer viruses appeared, so did antiviral systems. Computer viruses got cleverer and so did protection systems. Nowadays, nearly everyone works behind some kind of "firewall"—a computer version of the immune system.

Antiviral mechanisms exist at higher levels of complexity too, such as in the brain. We know less about cognitive "antiviral" mechanisms, but they too almost certainly exist. The brain is a tangled nest of interacting subsystems, each processing its own kind of information, and something must keep these in check and stop one of them from getting out of control. Sometimes this process fails—addiction, for example, arises when one of the motivational circuits becomes excessively strong and dominates behavior. We know little about the "cognitive immune system," but it is remarkably effective, given how well most of us function most of the time.

Logic tells us that antiviral processes must operate at the societal level too, because societies are themselves complex self-replicating information-processing systems. We are frequently reminded of this when we see them prey on each other, in the horrifying process that we call war, but even within a society we see frequent tensions as the self-replicating individuals that comprise the units of society battle for dominance. The same runaway propensity that plagues other complex self-replicating information systems also affects social interactions, with so-called "winner-take-all" dynamics in which the rich get richer, power begets power, etc., and societies become divided and unstable and sometimes eventually self-destruct. However, we also see processes that operate to obstruct these dynamics. Altruism is a not-yet-explained instinct that humans have to be generous to other unrelated people, including ones they do not even know. We experience this instinct as an urge to "be good" or "do the right thing," but it has long been a puzzle to evolutionary theorists as to how such apparently self-sacrificing behaviors managed to work their way into the genome, which tends to favor self-serving behaviors. The answer is, presumably, that altruistic behaviors *do* somehow favor the individuals that express them, although the details of how this happens are still being worked out.

There are other social behaviors that seem to have evolved to reign in the entropic forces of runaway antisocial self-interest. Societies often evolve rule systems—for example, both secular and religious laws—which regulate the behavior of individuals and make sure that they find a balance between self-interest and the needs of the communal whole. These "antiviral" systems are imperfect but have functioned well enough to allow 7.5 billion of us to rub shoulders reasonably well.

Putting this all together, then, we see that one thing that has allowed complex systems to persist has been the evolution of internal dynamics that detect and obstruct viral processes that would otherwise cause the system to become fatally infected and eventually self-destruct. Systems that fail to do this inevitably die. If we are to continue our spectacular technological progress, then, and make sure ours is one of the complex systems that persists, we need to learn from nature.

## Building Protections into AI

If quantum AI ever realizes the hype and becomes the super-fast, all-powerful boundary-pushing development that its proponents claim, there is a very real risk that we will be overrun by our AIs. This has sometimes been called the *technological singularity*—the point at which we lose control of our technology and it becomes self-sustaining and self-propagating. The physicist Stephen Hawking and tech entrepreneur Elon Musk have both warned, along with many others, that the technological singularity could result in human extinction, and that is broadly the thesis of this essay too. However, the lesson from four billion years of the evolution of complex systems on Earth is that sometimes systems evolve a way to cope with their explosive self-replicating dynamics, and the ones that do so are the ones that persist. Statistically, as one species out of the several billions that have ever evolved, our chances of averting extinction may seem small, especially given the technological singularity we are hurtling toward like a yacht toward a maelstrom. However, our intelligence, while the likely cause of our downfall, could also be the thing that prevents it, if we take our evolution into our own hands and learn from the lessons of the past.

This essay has laid out those lessons: we need to forearm ourselves with mechanisms for preventing the runaway self-propagation of unfettered self-replicating intelligent information systems. We need to make sure that alongside the development of these systems we develop constraints that prevent them from becoming harmfully autonomous—something equivalent to a DNA repair system or an immune system or a moral system. Quite how to achieve this is not yet clear, but our long-term success depends on figuring out a way. Entropy, which drives all complex-system development, is a race to the bottom, but the evolution of intelligence means that we have the potential to find a circuitous rather than direct route to that inevitable endpoint. Whether our AI designers are up to the task is . . . well, we'll see.

CHAPTER

# 14

# Quantum of Wisdom

Colin Allen, *Distinguished Professor in the Department of History and Philosophy of Science, University of Pittsburgh;* and Brett Karlan, *Postdoctoral Research Fellow in the Department of History and Philosophy of Science, University of Pittsburgh*

With IBM, Microsoft, Google, and Amazon all working to bring quantum computing to their cloud computing platforms, the projected benefits of quantum computing may seem poised to hit the mainstream. Yet numerous technical challenges remain for the development of quantum hardware and the algorithms to run on quantum computers. Physical hardware for quantum computing requires special methods for preserving the coherence of quantum states that are easily disturbed. The number of quantum bits ("qubits") in quantum computers has grown (on IBM's platform, for example, from 5 qubits in 2016 to

16 qubits at the end of 2020), and programmers have been given greater control over the topology within which the qubits interact. Yet, this is far short of the kind of exponential growth in computing power described by Moore's law, which has characterized computer engineering in the decades since microchips were first developed. In recent years, this growth has shown signs of slowing as physical limits to miniaturization of transistors have been approached. This has led some commentators to declare the end of Moore's law.[1] Others have pointed to quantum computing as the next technological development that will keep it going. However, both hardware design and the development of quantum algorithms to exploit the inherent parallelism of quantum states face tricky issues. Practical quantum computing thus remains speculative, and applications to artificial intelligence ("quantum AI") are even more speculative.

The authors of this essay are philosophers of cognitive science whose interest in AI (including machine learning) spans the use of AI for cognitive and scientific modeling, and the ethical impacts of deploying AI in various online and robotic applications. We are currently pursuing a project concerning how the mismatch between AI capacities and human understanding of those capacities presents barriers to wise use of the technology. Our particular focus in this article is on whether quantum computing presents any special issues for the ethics of AI. We are not concerned here with speculation about whether quantum effects are integral or essential to human intelligence or consciousness.

Despite the inherently speculative nature of quantum AI, questions about whether this future technology presents any special ethical issues are already beginning to take shape. As with any novel technology, one can be reasonably confident that the challenges presented by quantum AI will be a mixture of something new and something old. What little literature exists on this topic so far emphasizes continuity. For example, some argue that quantum computing does not substantially affect methods for achieving value alignment between AI and humans, although they allow further questions to arise concerning governance and verification of quantum AI applications.[2]

In this essay we turn our attention to the problem of identifying as-yet-unknown discontinuities that might result from quantum AI applications. Quantum mechanics is notoriously difficult to understand. (As Richard Feynman quipped, "Anyone who claims to understand quantum mechanics is either lying or crazy.") Insofar as quantum computing rests on some of the more mysterious aspects of quantum mechanics, and insofar as the various aspects of intelligence, whether natural or artificial, remain only dimly understood, quantum AI's position at the nexus of these presents novel challenges to its ethical use.

To understand the possible discontinuities and continuities of ethical questions in quantum AI, however, we first need a framework for thinking about ethical questions in AI in general. We are currently engaged in a project to frame issues of AI-human interaction in terms of *practical wisdom*. Our conception of practical wisdom with respect to technological artifacts has two main dimensions: broad and deep *knowledge* of the system's operating characteristics, on the one hand, and metacognitive awareness about the *limits* of that knowledge, on the other.[3] We believe that people engaged in various and often multiple roles—as designers, engineers, programmers, managers, salespeople, customers, and end users, as well as regulators and the public at large—need different combinations of understanding and support for developing the appropriate knowledge and metacognition required.

Numerous examples from AI already demonstrate the necessity of approaching problems with both a wide breadth of knowledge about the relevant conditions and metacognitive awareness of the shortcomings of that knowledge. State-of-the-art neural networks, for example, are vulnerable to adversarial attack, often via manipulations that are imperceptible to human perception. When using a computer vision algorithm to direct a self-driving car, how can programmers and users know when the car is likely to be fooled by a naturally occurring "edge case" that has not been seen during training or an adversarial attack introduced by another agent, or to make inaccurate judgments due to biases in its training data? Answering this question is complicated, but of paramount importance is knowledge of both the workings of the car and the exact instances in which blind spots and biases are likely to be consequential, or in which adversarial manipulation is likely to be encountered.

In the context of quantum computing, identifying the shortcomings of our knowledge is made significantly more difficult by the opacity inherent to the quantum system itself. The potential ethical pitfalls are nonetheless important to understand: if a self-driving car operated by a quantum computing architecture crashes and causes loss of human life, it might be physically impossible to recover information from the system without altering the system in unrecoverable ways. The ethical implications of this kind of opacity are multiple and worrying.

A practical wisdom perspective on the use of AI focuses stakeholders at all levels on anticipating new modes of failure in the technology. One can know many things about how an AI product will work without knowing the first thing about how, and in what ways, it will fail. Practical wisdom requires one to be aware of the limitations of an AI as well as its strengths. It also requires being aware of one's own limits in understanding the limitations of the technology. Increasing the practical wisdom of humans working with AI will require building sandboxes where stakeholders can evaluate the functioning of an AI within safe parameters, outside of real-world stakes and consequences of failure, and other spaces where people can brainstorm about potential problems and solutions without fear of repercussions. These sandboxes and spaces for discussion among multiple participants must be developed for all stages of product design, implementation, and deployment to ensure that wise design permeates all aspects of a project relevant to the largest numbers of stakeholders possible.

Providing sandboxes for limits-testing of new technologies at scale would present a particular challenge for any novel technology, given that such technology is always expensive and rare at the outset. The particularly technical aspects of quantum computing make it unlikely that it will become available in desktop devices any time soon. This means that access to quantum AI sandboxes will be more limited than for most technologies. Although quantum computing simulators that run on classical computers exist and may be adequate for programmers to prototype algorithms, they would be wholly unsuitable for the kinds of limits testing we envisage since practical knowledge of the performance characteristics of quantum AI for many real-world applications will crucially depend on having systems that respond quickly and thus depend crucially on the parallelism of true quantum computing. Such real-time responsiveness cannot be simulated; otherwise, the quantum computer would be unnecessary to begin with! This likely discontinuity with previous technologies requires an even more careful application of practical wisdom at upstream stages (design, planning, and implementation) to avoid possible catastrophic failures of products once they enter the mainstream, a situation made only more pressing by the prospect of limited access to the machines for the majority of their development and deployment.

The provision of testing sandboxes with limited access is, nevertheless, preferable to having no capacity for testing at all. Obviously, for any kind of AI, whether based on quantum computing or not, testing requires the technical expertise of experts in the relevant technologies. Many obstacles to wise use of AI, from algorithmic bias to data privacy, are problems that depend on, even if they are not fully solved by, technical solutions delivered by experts.[4] But limits testing of AI cannot be left solely to the experts. For one thing, the ethical and societal stakes of information technologies affect many more people than the experts at universities and technology companies, and those affected have a right to contribute to decisions. Additionally, it is often difficult for experts to predict how nonexperts will use new technologies or to predict how their behaviors will change (often in adverse ways) in response to new technologies.[5] It is thus crucial to develop workspaces and sandboxes where these unexpected outcomes can be identified and mitigated before the real-world launch of an AI product.

This requires a degree of interplay between experts and nonexperts, which may be particularly difficult to achieve for quantum AI, since there are often additional barriers to understanding quantum mechanics without specialized training. It is one thing (and surely no simple thing at that!) to ask a nonexpert to understand the complicated workings of a simple artificial neural network, realized on a classical computer. It is quite another to ask them to understand the workings of a quantum computer, something even experts struggle with. It is thus not at all obvious how one should design and use quantum computers in a way that encourages wise use among the broadest number of users.

But why is this kind of deep knowledge necessary for wise use? One possible reply to our proposal is that competent and wise car drivers don't need to understand the workings of internal combustion engines or their battery-driven alternatives. Operators should be able to get by using various heuristics that operate well in particular contexts without bringing the operator up to speed on the fundamental nature of quantum reality. We agree that, in general, utilizing such heuristics represents a rational, and often wise, way to manage one's limited cognitive resources.[6] But further knowledge becomes increasingly relevant at the limits. For instance, some knowledge of the difference between the two kinds of motor would be relevant to deciding which to rely upon in extremely dusty conditions. Likewise, the safest drivers are those who have driven cars beyond their limits in safe environments, have thus good knowledge of the conditions under which (for example) tires will lose traction, and have the ability to recognize when they are in a situation not covered by the training provided by their prior limits-testing experience. The same principles apply to the design and use of quantum computers: although we cannot expect all users to have PhDs in physics, having a sufficiently tested and wisely designed product obviates the need for deep thinking at the margins.

Wise use of quantum AI involves exploration of use conditions that are hard to predict in advance, and even harder to predict without some level of basic knowledge about how the hardware and algorithms work. For instance, one important issue for some quantum algorithms is that reading out the result of the system interferes with it in ways that are not recoverable. If a readout happens too soon or too late, the result may be incomplete or inaccurate, but the probabilistic nature of quantum mechanics causes uncertainty about when to terminate a computation.[7] Another example comes from the susceptibility of quantum computing to environmental noise and the lack of robust error correction that is possible. This represents a sharp discontinuity with present technology and presents a host of technical and ethical challenges for design: how can designers, companies, and regulatory entities be sure to manage the effects of a system that is unknowable in a way that current technology is not? We do not believe we know the answers to this question, nor do we think anyone could know them a priori. We do claim, however, that the framework of practical wisdom, and its focus on building breadth of knowledge and testing our knowledge at its limits, provides a uniquely suitable way of classifying, and attempting to solve, these problems.

Even if these problems can be handled reasonably under various conditions, the precise delineation of those conditions means that there are many "unknown unknowns" in quantum computing systems. This in turn makes it hard to have a good grip even on the limits of one's own understanding of those limits. In the context of AI, the differences between classical computing and quantum computing may be especially pernicious because of the tendency that many people have to anthropomorphically interpret the behavior and utterances of AI, supplying interpretations that attribute more intelligence than actually exists. More generally, the unintuitive role of nonclassical probability theory in quantum mechanics may create new kinds of edge cases for quantum AI that do not exist for classical computing. As is the nature of edge cases and limits, we are not in an epistemic position currently to know (and be able to mitigate) all of the threats to wise and ethical use that might arise at the margins of quantum computing. If we adopt a practical wisdom framework at all levels of design and implementation, however, we will be in a position to identify threats from edge cases and minimize them as much as we can when they do arise.

In summary, though we remain noncommittal on the long-term future for quantum AI and skeptical about its short-term significance, we believe that more analysis is both possible and desirable of the specific challenges that quantum AI presents to wise development. We do not pretend we know how this analysis will proceed; indeed, we think it is impossible to know the nature of many limits cases that might negatively impact individual lives and society at large before significant testing has been done. Our practical wisdom framework does, however, give one a way of approaching both the continuities and discontinuities in our ethical approaches to quantum computing. Some of those discontinuities rest on practical, but perhaps surmountable, limitations (e.g., the limited access many have to the hardware required to build quantum computers), while other discontinuities are inherent to the nature of the system itself (e.g., the inability to read the system without interfering). This is all the more reason, we think, to focus on precisely delineating and understanding the limits of our knowledge and building systems that take those limits into consideration when making important decisions about people's lives.

CHAPTER

# 15

# Human Imagination and HAL

Erik Viirre, *Professor in the Department of Neurosciences, University of California, San Diego*

The technology of quantum artificial intelligence brings forth a variety of issues, with a key question arising around what quantum AI will actually be used for. As the technology will be used on a variety of problems that ultimately depend on setting variables and their operational ranges, the very large question of expertise in programming a quantum computer and then managing their operations also comes to the fore. Who will have access to quantum computers and quantum AI, and who will have access to humans that will be essential to these machines?

Closely related to the setup of quantum AI operations will be assessment of their results. Quantum computing systems by their very nature are nondiscrete and thus will have "error ranges."

Further, incorrect settings of these systems will result in their exponential power giving exponential failures. Who will measure reliability? Most important, any internal "examination" by the quantum computing system of its computational results will not be available using current technology.

Given these uncertainties, what are the ideal means of ethical management of these systems and their outputs? Of course, we already deal with uncertain, indeterminate, unexaminable quantum AI systems: humans. Our own imagination operations show us that while we might evince *post-hoc* explanations of our ideas, we really do not know where ideas come from in the first place. Thus, our societal mechanisms for ethical management of human behavior is a good place to start when considering the ethical management of quantum AI. Perhaps most intriguing for us will be the management of a breakthrough: a conscious, imagining quantum AI, a technology not here . . . yet.

## What Problems Will Quantum AI Be Used For?

A crucial issue coming from the emergence of quantum AI is the nature of what tasks such a system might be enabled to act on in the near-term, once quantum computing is reliably and inexpensively implemented, and in the long-term on a theoretic "perfect" system. We (probably) won't send our email or balance our checkbooks on our quantum computer, so what might individuals or organizations use them for? In all likelihood, classical computers will be combined with quantum computing, both for inputting the data and accepting the outputs and for circumstances where substantial portions of a given problem or algorithm will be managed by the classical architecture with a soupçon of quantum computing at just the right time.[1]

Problems for quantum computing systems will be set in situations where some number of variables can be assigned a value between 0 and 1 and then have logic operations carried out on them, such as Grover's algorithm to carry out a database search. [2] Thus, problems with very large continuous variable datasets, such as predicting weather using previous weather reports or insurance analysis to predict losses, will likely be pioneering use cases. Economic "predictions" will of course be a huge industry, examining market fluctuations, opinions about governments, etc. Beyond this operational use of quantum computing, simulations of actual quantum mechanical processes, such as electronic, atomic, and molecular interactions, will be carried out. In chemistry, the prediction of chemical interactions will be made, especially the search for complex molecules like proteins in drug discovery.

Quantum AI will have two instantiations: optimization of AI processes and then use of the processes. Optimization of AI processes will make great use of quantum computing techniques, ironically, particularly on quantum computing–based problems themselves in many cases. Might a quantum computer examine itself for improved algorithms or even hardware evolutions? "Design" of problem solutions, by varying the classical logic approaches and isolating the elements best managed by quantum computing, will be a key AI technology. Of course, all of this technological development will be managed by good old *Homo sapiens*—at this stage there is little evidence or technological prognostication that quantum AI computing will be imbued with any new features, such as "understanding," let alone consciousness. Thus, the use of quantum AI computing will still be very much the realm of human operators, perhaps with some autonomy of optimization of algorithms by the quantum computing systems themselves, ultimately resulting in drastically more power for the humans to use on their problems.

## Variable Setting: Human Management

It is tragic that an essay about computing has to include garbage in, garbage out (GIGO), but this axiom is no less true with quantum AI than with any other computational device, perhaps even more so. All the trite aphorisms about computers screwing things up because of the human element are in operation here, but now *combined with both exponential power and results that will be unverifiable*.

Humans will still be required to encode the problems we want a quantum AI system to manage—indeed, such encoding will be somewhat of an arcane art and highly sought-after. Just stating problem variables for, say, an optimal path search or molecular force calculation will be largely beyond almost all humans. Quantum AI programmers may be both a limiting and a very expensive and scarce resource. Quantum computing system developers will need to build very sophisticated support systems for variable definition, and a substantial education program will be needed to teach programmers, operators, and user interpreters of results. What will we do with a result that seems definitive but has a substantial unreliability?

From an ethical-social-political point of view, this limiting human resource could be the major challenge with quantum AI: who are the programmers, and who controls them? The machines are currently expensive, but may well become ubiquitous if Moore's law continues to hold, doubling processing power every 18 months, as quantum computing performance improvement has generally done for the last 10 years. The humans are the key.

## Reliability of Quantum Computing

Is there some qualitative difference between results obtained by classical computing engines and the new quantum AI? The quantitative answer is exponential, but is there really a difference, say, between the use of a classical "supercomputer" and a legitimate quantum computer? Both types of systems are expensive to own and operate and require arcane, abstruse knowledge to design algorithms, program, prepare data, and interpret results. Ironically, classical computers may be more "reliable," in that the same computational request on any given dataset should reasonably give the same result, whereas quantum computing computations will always have a possibility of not operating or giving erroneous results. The mathematical class of problems that quantum computing operates on is in some ways the same as our classical 0–1 logic-based systems but are stupendously larger. Exponentially huge problems will be tackled with quantum computing and thus be available to quantum AI.

In a decoding problem or database search, the discrete answer may be the reliability check. If messages don't decode, the search was wrong. If you don't find the right bank account or property deed or other unique attribute, you know the system failed. Such problems would be "closed" having a unique answer, or none at all. However, many of the problems we mentioned earlier are more about likelihood rather than deterministic, yes/no answers; in other words, they are "open." How will we "know" if we got the "right" answer? Crucially, it may be common that rather than a unique "answer" to a problem, responses with some sort of reliability estimate will be provided. A "code breaker" decryption system might be useful only if the precise answer is provided, but a climate forecaster will have utility with some sort of range of answers, such as how many cloudy days or how much rain will fall. In planning for more non-discrete activities (for example, what different crops might we plant for a good harvest?), ranges will be acceptable.

The "collapse" of the set variables in a quantum computing problem will have two important features: the ability to very rapidly manage huge numbers of variables with very broad ranges of parameters, and a complete *in*ability to describe the origin and the reliability of the results of such collapse. Current quantum computing systems are highly unstable in their computational elements, a situation captured by the phrase "You never run a quantum computation on the same machine twice." Any individual computing element may or may not function normally. Thus running a "calculation" is actually quite difficult from a believing-the-answer point of view. Designers handle this by the expensive method of having multiple layers of redundancy and quality assurance techniques in their systems. "Programmers" of quantum computing systems must manage this reliability issue, which ironically means that even a rudimentary problem such as creating a truly "random" number may not actually be random on a given machine at a given time. By rerunning variable settings to verify the outcomes of quantum computations, one might establish some bounds of reliability, and one hopes that rapid, ubiquitous quantum computing resources will enable such reruns. Indeed, a critical feature of quantum computing offerings from various manufacturers will be inherent "reruns" that will demonstrate the reliability of results.

However, chaos theory suggests that even small changes in variable inputs (or indeed the management of variables by a system) might result in massively different overall results. Rain or shine, stocks up or down, toxic or nontoxic molecules: combining quantum computing with a large number of variables, the likelihood of inappropriate results may be very high. Again, human problem setters, programmers, and result interpreters will be very important. Might human intuition be useful yet? The quantum AI might play fantastic chess or Go, but will it "understand" the game or why it won or lost?

## Human Behavior as an Ethical Example

Arthur C. Clarke and Stanley Kubrick wanted to anticipate the future in their 1968 film, *2001: A Space Odyssey*. The central antagonist is the artificial intelligence, namely, the HAL 9000 **H**euristically Programmed **AL**gorithmic Computer. The humans are just first names really: Dave and Frank. HAL kills Frank and tries to kill Dave ostensibly in his "pursuit" of mission success. However, in the sequel *2010: Odyssey Two*, the story reveals that the human Jupiter mission managers of HAL lied to him in his "orders," resulting in logical contradictions and the eventual murder of his unsuspecting crewmembers.[3] It was only his original programmer, with skill and brilliance, who uncovered the contradictory instructions that led to madness and death. Storytelling aside, we surely should take to heart this fictional example to manage both the programming and operations of quantum intelligent systems and their designers and salespeople. Complexity managers will have to understand both the mechanisms of the computers and the reliability of the results. They will also have to understand the high priests who design, own, and operate the systems.

Human imagination is a place where we can look for analogies in the ethics of quantum AI operations. Any human can't really report where an assertion they make "comes from," although they can certainly tie up in a bow post-hoc descriptions of the things they say. And of course, in conversation between people, assertions about any agreed-upon decision can be dressed up with logic, statistics, and other means to demonstrate the reliability of the idea. Likewise, any quantum AI system's output will be made up of both statements of fact, logical propositions, and statistical reports, reports that will have to be tested. Akin to, say, the development of a drug or a technology for the safety of operations (a quantum AI autopilot, for example), the accuracy of the results will be balanced against their utility in a form of risk assessment.

Risk assessment is already the coin-of-the-realm of ethics managers, so while quantum AI will be a new technology, its ethical management will start with conventional procedures. "Informed consent" starts with descriptions of risks and benefits, but crucially also includes processes to manage the setup of situations where decisions with ethical implications are important: what-might-go-wrong events, recurring review, and public reporting of the procedures and results. Might we need representation by quantum AI on an ethics board? Will ethics board managers of quantum AI systems have "civilian" members whom the experts must inform and obtain approval from? What if civilians can't understand? What about unregulated quantum AI systems, commercially or privately or secretly government owned?

If human thinking and its ethical management is our example for quantum intelligence, a quantum imagination itself will be one breakthrough we might anticipate. What features of the quantum imagination will be important? Ironically, true introspection might be the most important: greater than the human mind. If a quantum imagination can be constructed that can definitively remark on its operations, reliability, and the origins of its outputs, might that be a more expansive consciousness than the human mind? Perhaps that will be a true machine mind and breakthrough.

CHAPTER

# 16

# A Critical Crossroad

Joseph N. Pelton, Dean Emeritus, *International Space University, and Founder, Arthur C. Clarke Foundation*

> "There is no security against the ultimate development of mechanical consciousness . . . Assume for the sake of argument that conscious beings have existed for some twenty million years: see what strides machines have made in the last thousand! May not the world last twenty million years longer? If so, what will they not in the end become?"
>
> —*Samuel Butler, 1871*

The long-term natural biological evolutionary process that proceeded via random genetic enhancement took hundreds of millions of years to evolve. The progression from alga to the brain power of a rat took hundreds of millions of years. It then took additional tens of millions of years to evolve from the capabilities of a rat's brain to that of a modern human. Indeed, the evolutionary journey from Lucy, the primitive Australopithecus being, to today's modern human intelligence apparently took more than three million years. But what are the consequences of a future driven by rapidly evolving "machine-driven" evolution?

The same speed of technological improvement does not require similar time constraints. Samuel Butler 150 years ago speculated on the power of what Ray Kurzweil calls the law of accelerating returns (LOAR) or the speed of compound accelerated growth. Today growth in technologies such as artificial intelligence, quantum computing, biological engineering, and more give rise to a host of concerns and opportunities. The 21$^{st}$ century is a time when the rate of acceleration in human knowledge is increasing—a fourth-order exponential that physicists call *jerk*. Indeed, we live in a time where humanity is being literally "jerked" in an unknown future.

Dr. Henry Markham, who heads the "Blue Brain" laboratory project, has followed a much more accelerated pathway. His team progressed from construction of one functioning neuron in his labs as of 2005 to an entire neocortical column (of 10,000 neurons) equivalent to a small part of a rat's brain in 2008. By 2011, his proto-brain development expanded to 100 columns, which he deemed a meso-circuit. By 2015, he had achieved the equivalent of a rat's brain with 100 of these functional meso-circuits. This functioning power of a rat's brain he equated to $10^{15}$ flops of processing speed and about $3 \times 10^{13}$ bytes of memory storage. Markham's current objective is to achieve a neuron functionality equivalent of a human brain by 2025. This would be artificial functional capacity equivalent to a memory storage capacity of $10^{17}$ bytes and computer processing speeds of $10^{18}$ flops.

In Markham's world, a human's brain is thus 1,000 times more capable than that of a rat's brain in both memory and neuron processing speeds. If Markham's team is successful, the artificial combination of electrical circuitry within his Blue Brain project will create a neuron functionality that is, in effect, a trillion times greater than when he started 20 years earlier. With the continued improvement of quantum computing systems, increasingly "smart" AI algorithms, and more and more capable robotic systems, the future rate of development in this field seemingly will only accelerate. The possible attainment of an "artificial electronic being" (AEB) with a high degree of semblance to an actual human brain seems to be on the brink of practical achievement.[1]

The conjunction of these types of technical capabilities that AI guru Ray Kurzweil has popularized in his writings as *The Singularity* may thus soon become a reality—perhaps in the 2025 to 2030 timeframe. As Kurzweil explains in his Law of Accelerating Returns, technological and biological evolution speeds up the rate of learning, abstraction, intelligence, and invention. Current rates of rapid innovation may in time even lead to development of a so-called "von Neumann" machine. This is a "machine" with artificial general intelligence, or a form of "self-awareness," that not only can allow the re-creation of itself but also improve its design and functionality as each re-fabrication process takes place.

Increasingly, these rapidly accelerating new capabilities will change life as it is lived today, and the ultimate result will inevitably be a dramatic change to human civilization. This change will not occur in a linear way, or even at what might be called a rapidly increasing speed. No, the rate of change will be exponential, and this is perhaps the greatest challenge to human civilization's future.

The truth is that human society is not well adapted to such a rapid rate of economic, social, political, and cultural shift that is inherent in a Law of Accelerating Returns world. Human social and political systems, especially democracies, respond to problems as they occur and thus are typically reactive and curative in nature rather than proactive. Digital processing and especially computing systems that operate at $10^{18}$ flops (Exabyte/second speeds) do not interface well with humans. Arthur C. Clarke described humans as carbon-based bipeds that typically process information at 64 kilobits/second speeds. Michael Crichton, in his book *Congo*, created a scene involving a U.S. Congressional hearing where a fictional General Martin told the assembled lawmakers that in the "next war" there would be many millions of weapon interactions a second and that only autonomous general artificial intelligence could make the needed responses—human reactions and decision-making would be much too slow. The sci-fi scene in Crichton's book two decades ago appears to be presenting itself as a quite real dilemma today. In 2016, an open letter from the Future of Life Institute and signed by many notables including Stephen Hawking and Elon Musk warned of the threats that they foresaw from the development of autonomous AI, especially when applied to weapon systems.[2]

The concerns that economically developed countries will face in the age of smart robotics, automated machines, and robots with intelligence will go well beyond the issue of autonomous AI making decisions about warfare. The issue of "super-automation" and technological unemployment or under-employment will likely be the first and most primal socio-economic issue to be addressed. There are, for instance, millions of truck drivers in the G-20 countries of the world that may be rendered obsolete by machine-intelligence and smart sensor systems, mounted on trucks and delivery vehicles. These systems will be able to drive trucks more safely and for 24 hours a day and not charge an hourly wage. AI algorithms that approximate human intelligence and use expert system programs will be able to operate machinery; engage in farming, mining, and forestry; and carry out the duties of many of today's service jobs.

The consequences of super-automation will go well beyond the issue of how to deal with technological unemployment. In most countries of the world, revenues needed to operate governments are based on income tax. Observers such as Bill Gates, author and former presidential candidate Andrew Wang, and the author of this essay have argued that as "smart machines" enter the labor force, their efforts should largely be assessed to pay a comparable form of "income tax" by the industries that, in effect, "hire" these automated workers. The tendency when change occurs is to look at the immediate first-order impacts such as loss of jobs and income and associated economic and social disruptions, but the point is that the disruptions that will come from super-automation will have secondary and tertiary impact.[3]

These types of disruptions were anticipated 50 years ago in Alvin Toffler's book *Future Shock* where he talked about high-tech and high-touch issues. At that time the focus was on the loss of manufacturing jobs.[4] Today, however, the disruptions are increasingly likely to affect the service jobs that constitute more than 80 percent of employment in economically developed countries. Over time, human labor has shifted along this trajectory:

1. Hunter/gatherer
2. Farmer, miner/craftsman
3. Manufacturing (in the age of the industrial revolution and manufacturing)
4. Services (in the post-industrial era)
5. The unknown? What people will be doing next (in the post-post-industrial age)

The problem is the discontinuities. The unknown represents a new age where machines are as smart as people—or likely much smarter. We cannot know for certain the precise nature of the profound economic, social, and cultural implications that will come with remarkable speed in the time of "Homo electronicus." What will our world look like when "smart machines" are increasingly creative and autonomous, do not require a wage, and are able to work 24 hours a day without cease?

One of the most fundamental issues will be the effect on military defense and warfare. Elon Musk has warned of the possibility of a future oppressive AI dictatorship where nonhuman intelligences rule humankind and hold the power of life and death.[5] The ultimate question of course is where and how might AEBs and humanity come into conflict or perhaps merge in some sort of intellectual mind-meld as envisioned in Arthur C. Clarke's *Childhood's End*. Many have envisioned how this might possibly be achieved by means of a bio-chip that would allow some sort of merged intelligence as one of the consequences of the "Singularity."

Scientist, philosopher, Jesuit priest, and polymath Pierre Teilhard de Chardin wrote in the *Phenomenon of Man* (1955) about the Noosphere and the Omega Point. He envisioned a possible future point of enlightenment in which all humanity might find a way to share knowledge. He called this realm the Noosphere. His writings have prompted many discussions about how his predictions of a globally shared knowledge might be achieved and how artificial intelligence might be merged with human thought. Such a bio-engineering achievement and the creation of such bio-chip technology, which allows the merging of human and machine intelligence, could transform the course of human history. This sort of "homo-bionic" being and merging of AI and human intelligence would presumably represent Teilhard's "Omega Stage."[6]

It is appealing to think of what the power of AI systems might bring in decades to come. These innovations might bring the world of clean energy systems or technology to cope with overpopulation, climate change, and zoonotic diseases and pandemics—a bright counterpoint to the dark possibilities of AI systems that could make life and death decisions in fighter jets and other weapons systems.

The nature of human invention seems almost always to be a one-way gate—once a technological invention is achieved, it is very hard to "uninvent" it. The idea of bio-chip interface between humanity and AEBs is hard to dismiss once conceived and may very well be a doorway into the future that readers of this essay may live to see.

The bringing to life of the world of AI inside of a human mind might also shine a light onto the future employment of humanity in the so-called post-post-industrial era. The world of the "Singularity," changed forever by the combined forces of AI and quantum computing, is today both exciting and at the same time forbidding. What is clear is that the Kurzweil's "Singularity" or Teilhard's Omega Stage is indeed near—perhaps as close as two decades away.

The technology that such an event will unlock and the rate at which it unlocks will be revolutionary. Its exponential advances will change modern society and especially democratic nations. Market capitalism appears little prepared for the changes that a quantum AI-driven world of innovation will bring, as outlined in my book *MegaCrunch*. Change will come throughout society, and the impact will be everywhere—in jobs and employment, in forms of governance and taxation, and in education, training, and culture. Countless social, economic, environmental, and technological disruptions will permeate the entire world. Issues such as sustainability and cultural preservation will collide with the very economic and social survival of human civilization as it is known today.

The challenges humanity will face within the bookends of the 21st century will be greater than ever before. If new capabilities such as AI cannot be uninvented, then the challenges and opportunities this life-altering technology presents must be faced and embraced with judgment, compassion, and intelligence for it is now—for better or worse—the human future.

CHAPTER

17

# Empathetic AI and Personalization Algorithms

Philippe Beaudoin, *CEO and Cofounder, Waverly;* and Alexander W. Butler, *Associate Director, Quantum Alliance Initiative, Hudson Institute*

In 1989 the World Wide Web was invented, and it came with a dream. Talking about this dream, Sir Tim Berners-Lee once wrote, "The web is more a social creation than a technical one. I designed it for a social effect—to help people work together—and not as a technical toy." Yet today, it looks like we're getting further and further away from the *togetherness* that Berners-Lee was dreaming about. Fake news is spreading like wildfire. People adopt radical positions and defend them vehemently. Anger dominates online discussions. The World Wide Web and the social media platforms it harbors are driving us apart.

Although we can see these divisive forces at play on social media, it's hard to fully understand the mechanisms that lead to them. Why do people write angry comments? Why do they reshare fake news? Why do they believe increasingly irrational claims?

Part of the answer lies in human psychology, but part of it also depends on what these platforms place in front of us. To understand why the web is driving us apart, it's useful to dive into the technologies that are being used to select the content we see.

Almost all technological platforms today have a similar objective: to offer people a personalized experience. In common parlance, we refer to the mechanisms that power such platforms as *algorithms*. Behind the scenes, however, these algorithms are complex systems involving traditional software, human oversight, and, increasingly, AI and machine learning.

Let's look under the hood of such an algorithm. To operate, the first thing they need is an objective. Something like: keeping the user on the platform for as long as possible, increasing the number of friends a user invites on the platform, or, more likely, a carefully crafted combination of such objectives. Choosing what the algorithm will aim for is solely in the hands of the platform provider. Large tech companies choose this objective very carefully, and the decision is typically made by high-ranking executives.

Algorithms also need a surface to act on, pixels they can play with. For social media, this is typically the user stream, which allows an algorithm to select the order in which content will be presented to the user.

The last but most important input to an algorithm is the user profile. This is any information about the user that can help the algorithm improve its objective. This is where algorithms have gone crazy over the years. Initial user profiles were quite simple: they looked at gender, the country of origin, the web browser they were using, and so on. Today, however, everything is fair game. Algorithms profile users by collecting the links on which they click, the time they spend on a web page, the speed at which they scroll on a mobile app, and countless other data points.

As AI and machine learning started playing an increasingly important role in algorithms, software engineers started to realize that the best way to help them achieve their objective is to feed them with as many profiling data points as possible. This led platform providers to aggregate immense collections of data points about users. The risk that such collections invade user privacy has often been pointed out. However, they pose another much more pernicious risk: that of allowing algorithms to achieve their objectives by exploiting every single, no matter how insignificant, cognitive flaw of the people who rely on them.

This exploitation of our cognitive flaws plays an important role in many of the negative impacts we trace back to social media. For example, if an algorithm is trying to increase the time someone spends on a platform, and if seeing an inflammatory post consistently causes that person to write a long angry reply, then the algorithm will learn that showing an inflammatory post is a good strategy for achieving its objective.

This can be alleviated by defining the objective more carefully. In the previous example, we could ask that the algorithm maximize the time a user spends on the platform while reducing the number of inflammatory posts they are presented with. However, given that this objective function is solely in the hands of the owners of the platform, users can't be certain that it is done with their best interests in mind—more importantly, it's obvious that users do not all share the same objectives.

So how do we get out of that conundrum? As is often the case, it appears that AI—which allowed this problem to become so widespread in the first place—could also offer part of the solution.

One area of AI that has made significant progress in recent years is natural language understanding (NLU). A system using NLU can process any piece of text in plain English (or any other language) and understand this text in a much more subtle way than was possible before. For example, if a text describes how to bake a cake, the system can process it and understand the required ingredients and the different steps needed to complete the recipe. If the text is an article, the system can understand the topics covered, the tone of the article, whether the treatment is deep or shallow, etc.

This offers a way out of the current approach: instead of building algorithms that profile users through countless data points and optimize for an opaque objective, we could instead let the user express for themselves, using the richness of natural language, how they would like their experience personalized. Users could write a text—let's call it a *living manifesto*—that acts like a recipe for their personalization algorithm.

A manifesto could allow a user to describe the topics that are important to them, the people they want to connect with, the ideas they are curious about, whether they prefer reading deeply personal content or shorter pieces of text, moments at which they like listening to a podcast, and so on. This constantly evolving document would be able to answer why someone prefers any given experience in any given context. Moreover, it would offer that answer in a way that is transparent to the user as they write the manifesto themselves. This is a stark departure from the current approach in which an AI is trying to guess what someone prefers using data points they are often unaware of and is precisely the type of algorithm that Waverly is developing now.[1]

A manifesto-based personalization algorithm no longer needs to collect any data about users. It simply needs to understand their manifesto. This has many benefits. First, it has a direct and positive impact on user privacy given that the manifesto is a fully transparent document. But the most important benefit is that, thanks to the manifesto, the user can consciously control their personalized experience. This makes it impossible for the personalization algorithm to exploit the user's unconscious cognitive flaws since these are not visible in the manifesto. For example, unless someone truly likes being angry, they are unlikely to ask for more inflammatory content in their manifesto.

In other words, where current algorithms try to guess what users prefer from data points that are not consciously emitted, a manifesto-based algorithm offers someone an opportunity to stop and think. It lets people express why they appreciate a given piece of content in a given context and has the potential to create a form of personalization that does not only reflect how someone behaves as they use a product but how they aspire to behave.

As opposed to traditional algorithms, however, a manifesto-based algorithm has one very important challenge. Traditional algorithms reach a personalized experience in a frictionless manner by accumulating data points generated in the normal flow of an application. On the other hand, a manifesto-based algorithm requires the user to do the heavy lifting of writing their manifesto. A difficult task.

Fortunately, this is a task to which the advancements of quantum computing offer significant promises of progress. With the development of quantum computing, complex problem solving, which has so far daunted even the most powerful supercomputers, becomes achievable. One such problem quantum computers will be especially adept at solving is that of artificial intelligence machine learning algorithms of the highest complexity. This emerging field of quantum AI can bridge the computational gap between our current AI and its associated problems and AI that is empathetic and conforms to society's ethical framework.

This is where future progress in AI will prove useful. We need to develop AI systems that can assist the user in coming up with a rich manifesto that reflects their deeper aspirations. We need to build an AI that can show empathy as it assists people, encouraging them to find words that capture who they aspire to be without unduly influencing them.

This is a tall order given that current chatbots aren't even that good at helping users accomplish well-defined tasks. That being said, there are reasons to be hopeful.

First, people can be prompted to think about difficult questions through relatively simple mechanisms. We could therefore imagine an empathetic AI not as a chatbot but as a system offering simple yet thoughtful activities that trigger such reflections.

Second, text synthesis is an area of AI that has known a lot of rapid progress in recent years. OpenAI GPT-3 is a great example of a system that manages to synthesize creative text that has the potential of sparking someone's imagination as they are writing their manifesto.

Third, the intersection of quantum technology and artificial intelligence must be explored, albeit carefully. To prevent the realization of the risks outlined previously in the field of quantum information sciences, society must develop an actionable and adaptive framework governing the development and deployment of ethical quantum computing and empathetic artificial intelligence. A key danger here is that, if left to develop without such an ethical framework, the unfettered development of quantum AI could potentially amplify the failings we see with classical AI, further entrench the negative aspects of social media we aim to mitigate, and potentially introduce new threat vectors in the fight against disinformation campaigns.

These are only some steps on the way to truly empathetic AI. Humans achieve empathy by being sensitive to an array of subtle signals: tone of voice, facial cues, posture, subtle choice of words . . . everything is a hint into someone's deeper thoughts. Empathy also requires an array of mechanisms to help others surface their thoughts without interfering with them. Asking the right question at the right time is a difficult art to master for humans, let alone for machines.

Building truly empathetic systems will undoubtedly require scientific advances. First, we're going to need the skills of many different scientific disciplines. Neuroscientists, psychologists, behavioral economists, and many others will have to put their knowledge and datasets together if we are going to make progress in empathetic AI. A more difficult and controversial aspect is whether the "spark of life" that allows humans to empathize with each other can be achieved through classical physics or if it will need us to venture within quantum mechanics. Some experts believe the latter to be true, which would mean that without the combination of quantum computing and artificial intelligence, truly empathetic AI agents are simply a nonstarter.

Importantly, there are risks associated with the development of empathetic AI. A system that helps you write down how you aspire to have your experience personalized is in a position to unduly influence you. We will therefore need to tread carefully as we develop such an AI. It's important, however, to notice that the status quo is plainly suboptimal, as current algorithms do not even bother trying to make people conscious of how their experience is controlled.

It is likewise important that society does not repeat the mistakes made in the development and deployment of classical AI, whereby the technology was developed and deployed without a universal understanding of nor international consensus on its ethical implications. Consequently, society now faces the dilemma of playing catchup and in some cases (such as empathetic AI) trying to rein in or even reverse the technology currently in the wild. The field of quantum computing is still in its infancy and yet provides us with a chance to proactively develop ethical standards for this next stage in emerging technology.

Just as society can learn from the failed ethical deployment of classical AI, we can likewise learn from the success of another historical technological leap. During the emergence of research into recombinant DNA in the 1970s, a group of scientists with sufficient foresight convened the Asilomar Conference on Recombinant DNA in 1975. The gathering of scientists and policy and legal experts alike successfully developed an actionable, ethical framework around the emerging issues of this technology and prevented the proliferation of biohazards and the development of malevolent biotechnology. Learning from this successful model, it is necessary that a similar international framework be defined for the development of ethical quantum technology, focusing on preventing a repetition of the mistakes already made. Rather than deploy a technology with the hope of it developing to conform to society's ethical and empathetic standards, in the words of the Asilomar Conference's organizer, "It would be more effective, especially in the face of uncertainty, to provide guidelines that will undergo timely changes in response to new scientific knowledge."

It is crucial that we learn from history—not only of the failings of classical AI in this realm, but from the success of other ethical standards for similar novel technology—and develop ethical and empathetic quantum artificial intelligence.

Decades ago, Sir Tim Berners-Lee dreamt of building the World Wide Web as a network that would help people work more closely together. If we are to achieve that, we will need to fight the growing divisive forces of online platforms. Part of the solution lies in the development of alternative ways to personalize people's experiences. We'll need to free ourselves from the current generation of algorithms and their ability to exploit our cognitive flaws. These alternative systems will require the development of empathetic AIs that can understand us at a deeper level and that can help us express our aspirations without influencing them. Developing such systems will not be easy. We may even need to cautiously dive into the world of quantum artificial intelligence. The challenge shouldn't stop us from trying, though, and there are clear steps that can be taken today and that could pave the way toward a much brighter future in which people can truly come together. Quantum AI will provide us with the opportunity to fundamentally change the trajectory of our online behaviors and help us realize Berners-Lee's vision.

CHAPTER

# 18

# Should We Let the Machine Decide What Is Meaningful?

J. M. Taylor, *Co-director, Joint Center for Quantum Information and Computer Science, University of Maryland*

Building a quantum computer that does something well beyond classical computing, such as factoring a large number, is stupendously hard. So hard, I would argue, that it will lead to substantial changes in our computational and engineering infrastructure to achieve it, including meaningful advances in artificial intelligence subsystems. Only with these advances will we approach the regime of millions of quantum bits. Along the way, we can hope to promote AI advances that build in the necessary elements to ensure their benefit for humanity.

The central question for discussion here is thus: given that we will rely ever increasingly upon machines to build quantum computers, including machine learning and more complex artificial intelligence systems, will we be able to even understand the consequences of the systems we have created? Or will our understanding be available only through the filters that our machinery lets us observe?

But first, how hard is building a large-scale quantum computer? To give some context, consider the famous Schrödinger's cat thought experiment. In this experiment, Schrödinger posed the question: if a cat is put in a box with a qubit connected to a vial of poison gas such that the poison is released if the qubit is up but kept in the vial if the qubit is down, what happens when the qubit is set in superposition of up and down and the experiment run? He contended it was nonsense, largely because we could not build a box that was isolated enough for the superposition of the qubit to lead to the entangled state of the cat alive and dead at the same time.

More recently, researchers showed that being able to measure the existence of this superposition requires the capability to reverse the physical operation leading to the dead cat.[1] That is, to prove to another observer that you have created a Schrödinger cat, you must be able to control the cat+poison+qubit system sufficiently well to be able to reverse all the operations used. This so-called quantum necromancy is mostly a thought experiment still. However, it highlights the challenge of building a large-scale quantum computer. One must develop the tools for the recovery of information to the point of keeping a superposition of tens of thousands, or millions, of qubits. That is, we must be able to completely reverse (in principle) all that has transpired from the beginning of the computation. That is still much easier than making a dead cat live. Nonetheless, it does push the limits of what we know how to achieve today—to wit, our engineering capabilities. Indeed, it is already necessitating advances in fields from materials science to microwave engineering to computer science and more.

We now turn to the benefits of investing in quantum computing. I believe that achieving this crazy goal is, in its own right, worthy as it tells us whether the universe computes quantum mechanically, which echoes beliefs of others in the field such as Krysta Svore, Scott Aaronson, and John Preskill. But, it also turns out that investing in how and what we can compute continues to have tremendous returns for humanity and our capabilities as individuals, nations, and the world. Specifically, the technology advances necessary have many knock-on questions. Indeed, the gains in productivity of the past 50 years have been largely driven by the capabilities enabled in new computation and networking technologies and applications. We take this for granted today, but the ease of thousands of transactions in our typical week is a stark contrast to the same set of transactions from 50 years ago, from paying bills to ordering groceries to making operational decisions in a complex supply chain. This reduction in unpleasantness in our lives, to take a phrase from Robert Gordon, has been a boon to our society. What future boons await?

## What Is the Future of Computing?

Thus, quantum computing is not only about quantum necromancy, but also about enabling the future of the human race. It is a particular future of computing, one that sits alongside other novel computing paradigms that are likely to complement traditional computing approaches (today's silicon chips operating with extremely low error rates). A key aspect of these future computing systems is their greater capacity to allow for errors in the computation itself.

For example, neuromorphic, analog, and approximate computing systems can operate at much lower energy per operation, at the price of producing noisy, or sometimes different, results when run with the same inputs. Why would we accept these outcomes? Part of it returns to the quote of Alexander Pope: "To err is human; to forgive, divine." Specifically, dealing with various artificial intelligence subsystems researchers and companies have found that, in practice, approximate results are sufficient for the vast majority of cases.

One way to understand this is to examine how the outcomes of the training process (a type of optimization) depend upon thousands or even millions of examples for the systems to learn from. The landscape of performance cannot depend substantially upon any single one of those data points, and thus we find that changing some of the parameters of the neural network circuit a small amount leads to only small changes in the output. This lack of change is sometimes called the *barren plateau problem*, as it makes it difficult to train systems past a certain point. But this curse is also a blessing: it means small errors do not contribute meaningfully to the overall performance of the system. These errors reflect the approximate nature of the human reality around us.

Nonetheless, these paradigms, combined with advances in algorithms, are taking classical computing to new places already. And as presaged before, it is also reducing our ability to understand the specifics of the systems we have created—the data is, in many respects, the code, and the data is vast.

## What AI Is Enabled by Future Computing Paradigms?

Specifically, as the future of computing is evolving, there are new applications and algorithms emerging that represent the beginning of what many term *artificial intelligence* in a more general setting. That said, we will stay away from "general" AI and instead examine the disparate set of techniques that fall under the AI rubric, which all connect to aspects of our human experience of "intelligence."

Some of these techniques are motivated by our understanding of how organisms perceive the world. They correspond to elements such as recognition of images, text, sounds, and other sensory input. They use concepts drawn from how neurons in an animal behave to develop fuzzy estimates of what a set of inputs corresponds to, given sufficient training. They amplify and inform us about the world around us in real time and enable the automation of many menial tasks.

However, these systems already are beset with challenges, particularly around the difficulty of bias and truth. When you use automated recognition systems, you are delegating questions of the basic truth to both the dataset used for training or execution and to the assumptions (hidden and known) in the AI-related subsystem used. These lead to sometimes humorous but often horrendous results, from profiling that highlights white, male faces over others to misunderstandings of cause and effect.

More complex AI subsystems delve more deeply into emulating or complementing the trickier recesses of thinking and the mind. They look at enabling augmented or artificial creative thought, learning, and decision-making. Here the mathematical models and the approaches diverge from the biological networks, as our scientific knowledge of brain function does not have sufficient understanding of these systems to enable a reverse-engineering approach (unlike in simple neural networks). And again, fraught challenges confront us, as we have to wrestle with what decision-making can be delegated to automated systems, and at what cost. We also must recognize that the Pandora's box of computer-driven exploration may reveal more techniques and technology than we, as a species, are capable of easily integrating and responsibly using.

These topics are very difficult to build a consensus on, and I will not detail my thinking on them here. My key statement is simple: I believe that it is exactly our humanity that allows us to choose to use these systems for the benefit of our society and civilization. When we lose sight of why we use technology and instead blindly enable world-changing technologies to take over our lives, we are abdicating our individual and collective moral responsibility. Thus, we must develop these technologies for purposes that are tailored for their effective execution and, limited to that arena, much like a sandbox in information security.

## An Illustration of Developing AI Subsystems from Physics

To illustrate this point, I would like to focus on a smaller story to showcase how keeping humans in the center of the picture enables new regimes of technological execution.

At the end of 2020, a landmark scientific paper[2] was published in *Science* magazine, showcasing how a quantum device can execute a very specific algorithm exponentially faster than any classical computer, a topic I wrote on three years ago.[3] This was not the first paper demonstrating this particular step forward (Google did it a year prior).[4] However, it was done using light rather than circuits, it was done with a system that has been considered a "dark horse" candidate for quantum computing, and perhaps most interestingly it relies much more heavily on human labor. It is also the first landmark result in quantum computing for a China-based research group.

While there are a variety of intriguing ways to examine the result and its meaning, here I'd like to explore the role automation plays in physics experiments and connect it to emerging uses of automation in other areas of our lives. With the advancing discussions around us about machine learning and autonomous vehicles, one starts to wonder how dystopian our science fiction future may become.

The work from Chao-Yang Lu's group described in the *Science* paper highlights the interplay effectively: there is a huge amount of manual labor necessary to realize this result. If you examine the photograph from the group, you see the tremendous number of hand-aligned optical elements, each of which needs to point with micrometer precision at the right things. For me, this reflects a greater truth about most cutting-edge physics experiments.

Specifically, in physics experiments (and other fields as well, but I am not an expert in automation in biology, for example), the next experiment you do typically builds from your last successful experiment. Usually, you look at the apparatus and ascertain which elements you might successfully automate so that the humans running the experiment can focus on something more important. You do this until you are no longer able to automate; then you try the experiment and ideally succeed. I have the sense, working closely with cutting-edge teams, that the limits of our knowledge are set by the rate at which we can implement this loop and understand the consequences of the increasing automation.

For example, many groups working on building quantum computers around the world focused on how to better leverage automation. This included hardware automation, such as using nanofabrication and lithography techniques to build the complex interference systems. It also included substantial software automation in testing systems, tuning systems, and running systems, which may enable their effort to scale more rapidly than the recent Lu group results. As the CEO of Honeywell recently noted, their main value addition to the space of ion trap quantum computing is not cutting-edge physics *per se*, but rather their expertise in control and automation of complex systems.

## Takeaways for the Incorporation of AI Subsystems into Society

What can we learn from the decision process for automating part of an experimental setup? I would characterize it as driven primarily by a recognition that repetition of a job is something ideally relegated to automated systems. Usually, the decision to automate rests with the researcher actually doing the day-to-day automated work and thus has a high incentive to find a more efficient use of their time. Thus the first step is **having the affected community develop the initial solution**.

However, the process only starts there. Once an opportunity for automation is identified and a nascent solution prototyped, there are a variety of checks and improvements used to ensure a quality final product, namely, a reliable device for doing science. This includes other members of the lab or research team checking the work; running the system through unit tests and other means of confirming it performs within parameters and does not, for example, break the system. This takes the form of both software and hardware *sandboxing*: **systems should have limits designated by humans to ensure performance in the first instance does minimal harm**.

As an example of how this works, when I was a graduate student, I wrote a feedback routine for stabilizing a large superconducting magnet in a dilution refrigerator—a system that operates a hundredth of a degree above absolute zero. Unfortunately, I had a sign error in the feedback loop, and when implemented, the magnet promptly ramped up to its maximum current. Fortunately, I had a "heartbeat" code that prevented changes in the apparatus faster than the measurement system could keep up, and this prevented an accidental quench of the system and the attending venting of all the liquid helium into the room. Thus, the sandbox was limited to prevent the accidental behavior. Such engineering controls are essential.

Finally, the system must yield results that advance the goals of society! For quantum computing, this means that the scientists shepherd the application and the continued evolution of the automation systems, with checks on performance and outputs. These checks are not just statistical, but also intuitive. Thus, stakeholders with expertise must **continually challenge the automation systems to prove they are trending toward the intuition and the hard data expected**.

These three principles form a natural core for the application of machine learning and artificial intelligence systems to problems relevant to humanity, from self-driving vehicles to new ways of discovering information and creating art. They reflect a humanist viewpoint: these systems exist to enable us to flourish, individually and as a species. Critically, they necessarily shy away from the dystopian future found in novels like Neal Stephenson's *Snowcrash*, where a system much like an advanced AI subsystem effectively hijacks the brain's decision circuits, the science fiction future of behavioral economics and marketing on steroids. To avoid those outcomes, **affected stakeholders must be part- or full-time active developers**. This means that education is the implicit last component for enabling positive progress.

## Some Steps Forward

One part of this approach that speaks to me is the key connection between the technology, the teams that implement the technology, and the users of the technology. This is why I look so closely at public-purpose consortia[5] as a means of keeping these communities connected and growing together in concert.

To summarize potential pathways forward for quantum AI:

- Community-driven automation puts humans at the key decision points for the creation of a new system.
- Sandboxing, including both software limits and engineering controls (interlocks that protect us from our own stupidity and from the "unknown unknowns"), comes at the beginning, middle, and end of the projects.
- Automated systems application and performance require stewardship, with experts and stakeholders continually challenging their outcomes and expectations and testing them against both data and intuition.
- Communities impacted by automation should be part of the development cycle, including in their education and in their deployment.
- Mission-specific applications are a critical way to ensure that the interplay is maintained and sustained throughout the development of AI subsystems and will accelerate the development of new machine learning and AI-related techniques.

Even with these small steps, there are great problems that remain. In the words of my student Shangjie Guo, "When a human is unable to process the data, or the statistics (or even structure) of the data, should we let the machine decide what is meaningful?" Or, what is the delegation that we can enable and will allow? Separately, what about bias and understanding of decision-making—our ability to understand the basis and process of a decision or recognition task? These, and myriad others, await us; only by containing explicit systems to well-defined domains (sandboxing) can we progress the technological foundation while this additional critical work is done.

These steps we already recognize as essential for advancing the forefront of human knowledge. I have a great hope that they can also be helpful in enabling humanity to use our time more effectively while not sacrificing the essential elements of our society, culture, and life.

CHAPTER

# 19

# The Ascent of Quantum Intelligence in Steiner's Age of the Consciousness Soul

Stephen R. Waite, *Author and Adjunct Scholar, Quantum Alliance Initiative, Hudson Institute*

When thinking about quantum computing, artificial intelligence, and the technological landscape of the future, many scientists and futurists express genuine concern about machines overtaking humanity. Just the other day, a YouTube video popped up on the screen with the title *Will A.I. Kill Us*, with a discussion of this question by Dr. Ben Goertzel. Screenwriters in Hollywood have penned many scripts with powerful AI machines posing a grave threat to humanity.

In considering the potential societal and economic impacts of quantum computers and AI of the future, it is necessary to consider the state of consciousness of humanity. Conventional science doesn't have much to say about the subject of consciousness today, but I believe it is vital to the future of humanity and is worthy of investigation.

Consciousness is the great mystery of science. In his book, *The Quantum Revelation*, Paul Levy notes that physics is encountering consciousness and is only in the beginning stages of consciously realizing it. In essence, says Levy, consciousness has entered into the physics laboratory, and physicists are not quite sure what to make of this turn of events. Levy goes on to state that most physicists think that something as ethereal as consciousness has no place in "real" physics.

The father of quantum mechanics, Max Plank, appeared to be ahead of his fellow physicists. It was Plank who, in 1931, said, "I regard consciousness as fundamental. I regard matter as derivative from consciousness. We cannot get behind consciousness. Everything that we talk about, everything that we regard as existing, postulates consciousness."

It was quantum physics that revealed the proverbial elephant in the room that has quietly been swept aside by scientists over the decades. The elephant is the effect that observing has on the results of scientific experiments. The famous double-slit experiment demonstrates that the very act of observation can alter the results. The peculiar nature of the quantum realm opens the door to the possibility that consciousness and the material world interact in fundamental ways, something that helps to put Plank's statement on consciousness into context: everything scientists talk about and regard as existing posits consciousness.

Another prominent scientist who talked a great deal about consciousness during the ascent of quantum mechanics in the early 20th century was Dr. Rudolf Steiner. Dr. Steiner is not as well-known as Planck, and there are quite a few mainstream scientists who are unfamiliar with his work.

As it turns out, Steiner had much to say about consciousness and its role in humanity when others had little or nothing to say about the subject. In fact, much of what Steiner wrote and spoke about was based on an evolution of human consciousness. In hindsight, he appears to have been ahead of his time—a scientific prophet, as it were.

Steiner delineated history into different epochs. Each epoch was characterized by a state of consciousness, and he described the current period as the age of the "Consciousness Soul." This is not the place for a deep dive into Steiner's view of the human being and the evolution of consciousness. What is sufficient for our purposes is to note that the age of the Consciousness Soul represents a departure from the previous state of consciousness. The previous age was what Steiner referred to as the "Intellectual Soul" (sometimes referred to as the "Mind Soul"). The Intellectual or Mind Soul was entangled in sensations, drives, emotions and so forth. It was during this age that Newtonian mechanics and materialistic science manifested. The workings of the cosmos became viewed as a great machine. Quantum mechanics emerged after the end of the Intellectual Soul period and the beginning of the Consciousness Soul era.

Steiner believed that the age of the Consciousness Soul will bring with it a movement away from materialism and materialistic science. In the age of the Consciousness Soul, human beings will increasingly allow what is true and good to come to life within us. Steiner said that what the soul carries within itself as truth and goodness is immortal. The eternal element that lights up within human souls, says Steiner, is the Consciousness Soul.

As human consciousness evolves in the years and decades ahead and the light of truth and goodness comes to life within a greater share of humanity, we are likely to see an evolution away from materialistic views of nature and the universe. The whole fabric of science is likely to change, perhaps as dramatically from the evolution of Newtonian or classic mechanics to quantum mechanics.

After studying quantum mechanics, Sir James Jeans remarked that the stream of human knowledge is heading toward a nonmechanical reality. Through a quantum science lens, said Jeans, the universe begins to look more like a great thought than a great machine. As human consciousness evolves in the decades ahead, quantum science and science in general—aided by quantum computers—is likely to begin to study nonphysical phenomena more seriously than it has in the past. As we know, materialism and materialistic thought and science flowered during the Newtonian age. The quantum age is one filled with nonmaterialistic potential.

It was the great 20th century inventor Nikola Tesla who proclaimed during the ascent of quantum mechanics, "The day science begins to study nonphysical phenomena, it will make more progress in one decade than in all the previous centuries of its existence." That's a bold statement to be sure, but one that is not likely to be far off the mark.

We can already detect a shift in the direction of science toward nonphysical phenomena. In May 2018, the *American Psychologist*—the flagship journal of the American Psychological Association (APA), the principal U.S.-based professional organization for clinical and academic psychologists—published a lead article titled "The Experimental Evidence for Parapsychological Phenomena: A Review." The article's author, Etzel Cardena, professor of psychology at Lund University in Sweden, analyzed 10 classes of experiments exploring psychic "psi" effects.

Cardena found that the evidence for psi is comparable to that for established phenomena in psychology and other disciplines. It is noteworthy that this article appears in what is considered a conservative voice of academic psychology—a scientific discipline that has been extremely skeptical about psychic abilities. As psi researcher and author Dean Radin notes, the publication of Cardena's article represents an academic sea change in subjects that are appropriate for serious debate.

With the rise of quantum computers comes the possibility to study nonphysical phenomena in ways that scientists could never do previously. The study of superconductivity alone using quantum computers, in part powered by superconducting quantum processors, could take science into a whole other realm of investigation. Things that have baffled scientists over the decades could become explainable through the use of quantum computers and the computations they enable.

But more importantly—and this is a critical point—the evolution of human consciousness holds the key to the economic and societal benefits of quantum computing, AI, and other advanced technologies that manifest in the future. As Nobel laureate and quantum computing advocate Richard Feynman noted, technology is a double-edged sword: it can be a blessing to humanity, but it can also be a curse. Should humanity evolve to higher levels of consciousness in the age of the Consciousness Soul—and subsequent ages to come, as Steiner foresaw—one would hope that advanced technology will be used more to benefit humanity instead of curse it.

Should humanity evolve in the manner envisioned by Rudolf Steiner—that is, toward higher levels of consciousness—the use of AI and quantum technologies may shift toward the greater benefit of humanity. As futurist George Gilder notes, AI will work best when it is devoted to enhancing human minds rather than trying to usurp or replace them.

The great science fiction author and scientist Isaac Asimov envisioned such a future for AI in his "Bicentennial Man," which possessed a far higher level of consciousness than any technology today. Asimov's AI technology obeyed several laws, one of which was "A robot may not harm a human being." It was Asimov who envisioned AI that could love. Scientists today have no idea how to program a machine that loves. In fact, scientists have little to no understanding today about the nature of love at all.

Asimov's imagined futuristic machines were capable of loving and doing no harm. Such enlightened machines undoubtedly would be a tremendous blessing to humanity. They would reflect, among other things, a more advanced state of humanity where love consciousness is far more pervasive on Earth. The human brain has served as inspiration for the current generation of AI, but when coupled with the quantum computers of the future, AI could begin to manifest characteristics of the human heart, radiate love, and begin to resemble Asimov's Bicentennial Man.

I firmly believe that it will take a further evolution of human consciousness to realize the full benefits of quantum computing for all humanity. The role of human consciousness in the evolution of technology should not be underestimated. It is, in fact, crucial to how advanced technologies will be used in the future. This cannot be overstated.

## Into the Quantum Age . . . and Beyond

The manifestation of carbon-based quantum computers has the potential to manifest intelligence that far exceeds any AI today. It is here where the potential for real, global economic prosperity lies. Imagine a quantum computer with exponential power and capability that has the potential to overthrow the chains of scarcity that have long dominated economic analysis and usher in an age of abundance economists can only dream of today.

One thing is fairly certain. Quantum computers and the knowledge and innovations they help manifest have the potential to be a far bigger boon to society at large if programmed and used by human beings possessing elevated levels of consciousness. After all, the evolution of human consciousness toward higher levels, not quantum technologies, is likely to be a far more important determinant of the health and well-being of humanity in the future.

Physicists could learn a great deal from the study of how enlightened human beings (e.g., mystics, sages) see the cosmos—not in terms of religion, but in scientific terms and as a way to foster the advancement of science and knowledge, in general. Quantum physicist David Bohm was working along such a path in the previous century with the mystic Jiddu Krishnamurti, but few have seemed to follow his lead. Paramahansa Yogananda—of whom Apple cofounder Steve Jobs was a follower—was keen to bring the views of enlightened human beings into science as well.

Manifesting a world of economic abundance will not be easy. It will require an upward shift in human consciousness far beyond apathy and fear. An elevation in human consciousness, at a time of growing machine intelligence, would help maximize achieving the full benefits of what quantum computing can offer humanity in the future. In the concluding paragraph of his book *Love and Math: The Hidden Heart of Reality*, Edward Frenkel stated:

> We are pondering eternal questions of truth and beauty. And the more we learn about mathematics—this magic hidden universe—the more we realize how little we know, how much more lies ahead. Our journey continues.

As Albert Einstein saw, scientists must remain humble no matter what the age. Retaining an open mind is essential to scientific progress and the evolution of human consciousness. It is also worth noting there is no role for propaganda and censoring in scientific pursuits, nor is there a role for the politicization of science. Freedom and humility, not propaganda and censoring, are the things that will lead humanity onward into the quantum age.

There is a vast unexplored nonphysical frontier that quantum computers and other quantum technologies could help illuminate in the future. Such illumination may well hold the key to understanding the essence of consciousness and the many mysteries that have intrigued and puzzled scientists for ages. As Einstein reminds us:

> Everyone who is seriously involved in the pursuit of science becomes convinced that some spirit is manifest in the laws of the universe—a spirit vastly superior to that of man, and one in the face of which we with our modest powers must feel humble.

As the quantum age evolves, let us always try to feel humble.

CHAPTER

# 20

# Quantum Computing's Beautiful Accidents

Christopher Savoie, *Founder and CEO, Zapata Computing*

"Beauty is the ultimate defense against complexity."

—David Gelernter

On an early November day in 1895, Wilhelm Röntgen, professor of Physics in Wurzburg, Bavaria, altered the future of medicine for the better. But not because of what he originally set out to do. In fact, what he *actually* discovered did not match what he *wanted* to discover.

While ascertaining whether cathode rays could pass through glass, he noticed an incandescent green light seeping through the black paper-covered tube and projecting onto a fluorescent screen. Intrigued, he continued experimenting to learn more about this strange green light. He eventually found that it could pass through different substances and leave behind visible shadows of solid objects, including human bones. Not knowing what they were, Röntgen called them *X-rays*. These discoveries eventually changed all of medicine in a profoundly positive—albeit negligently arrived at—way.

I couldn't help but think of Röntgen's breakthrough when the opportunity to be part of this book came along. The discovery of X-rays is only one example of a trend I like to call *negligently positive outcomes*. Penicillin, Velcro, vulcanized rubber, and insulin are a few other well-known cases of this phenomenon. And, depending on your age, you can throw computers begetting a thriving global gaming industry into the mix. As a scientist myself, I contemplated that this will inevitably be the case for quantum computing as well when, together with five other scientists, we set out to found a quantum computing software company. Indeed, this process of unintentional invention is an intrinsic and vital aspect to technological and scientific innovation. Frankly, the world is so complex that it's impossible to truly know all the unintentional consequences of quite intentional actions, big and small.

Complexity is all relative, of course. Finding your way through Boston as a tourist without a map or GPS guide is one level of complexity (and not a small one!). Figuring out how to take advantage of quantum computing's power to speed up drug discovery is at another level altogether.

Every day people around the world attempt complex undertakings to varying degrees of success. In the course of navigating this complexity, they negligently generate both positive and negative outcomes as a result. The difference with quantum computing is that the technology is exponential by its nature and brings a—excuse the intentional pun—quantum leap in technological capabilities that even the experts do not fully understand the limits of. Quantum computation has the great potential to revolutionize business and life in myriad ways, and at a level that civilization has yet to experience. So, yes, a negligent outcome could be life-altering on a massive scale (not that getting lost in Boston's maze of winding one-way streets isn't a life-altering, albeit common, negative event in the current pre-quantum era).

## Unlocking Human and Business Value, One Way or the Other

In the next decade—even the next several years—we will likely see quantum computing's impact unfold first in the fields of artificial intelligence (AI) and machine learning (ML). The advantages will manifest themselves in how we optimize our resources, in how we grow our food, in how we predict complex events like climate change and the spread of pandemics. Such systems are obviously too complex to model by the human mind. Alas, they are beyond the scope of accurate prediction by even the most advanced classical supercomputers currently available, even with the tremendous strides we have made of late in machine learning and artificial intelligence. Case in point, the COVID-19 pandemic—what would the world not have given to have more accurate models of the epidemiology and spread of this virus as the pandemic unfolded. Many unfortunate policy decisions were made due to the lack of accurate scientific models of viral spread and models of disruption in global supply chains and the economy that ensued. Good policy can be made only with good data and accurate models. In relatively short order, quantum computing will deliver modeling of complex systems that has been, until now, beyond the reach of our most advanced technologies.

In the pursuit of these outcomes, my colleagues and I intend to build quantum computing applications that generate positive outcomes and unlock a huge amount of human and business value. The key word here is *intend*. While we have already begun to apply quantum computing to create intentional positive outcomes—such as in financial portfolio optimization, chemical modeling for drug discovery, and generative adversarial networks (GANs) that have the capability of creating accurate synthetic data beyond the capabilities of classical computers—additional negligently positive outcomes will also be generated along the way that nobody today could possibly foresee.

I would be remiss if I didn't mention that, as with any emergent technology, some negative outcomes are also possible with quantum computing. Whether intentional (e.g., breaking encryption) or negligent and unintentional (e.g., new privacy concerns brought about by a capability to accurately predict human behavior, perhaps such as was foreseen by the popular TV drama *Person of Interest*), negligent outcomes tend to be a double-edged sword.

There is no need to overly obsess on dystopian outcomes, but others certainly have crossed into the territory of cautionary hysteria. This is why my colleagues and I strongly believe that quantum computing needs governance and guidance. The same can be said of AI and machine learning, and there are, fortunately, organizations and individuals pursuing this path. For example, recently some fellow attorney colleagues and I, under the auspices of the American Bar Association, formed a task force to explore the societal, legal, and policy implications of quantum computing as it emerges as a force of societal change.

I am a career entrepreneur at heart and therefore an optimist by nature and as such am personally drawn much more to the negligently *positive* outcomes since they are perhaps the most beautiful and inspiring. That's why the potential of quantum computing is so tantalizing. By using quantum thinking to rewire how we think about problems in a more abundant, probabilistic, scenario-driven way, we will create negligently positive outcomes we could not have anticipated otherwise.

But first things first.

## Quantum as a Force Multiplier for AI

Though quantum computing continues to evolve, the actual devices are not yet faster than their classical counterparts. In fact, if you saw one of them, it would probably remind you of a giant, futuristic train set—a supercooled, near-zero degrees Kelvin one, that is. Some say it reminds them of the beginning of modern computing as exemplified by the room-sized ENIAC machine.

Depending on who you're talking to, quantum computing won't come into its own until there's a large-scale, fault-tolerant computer. Is that five years from now? Ten? More? It's impossible to say. But I would counter that argument by pointing to the existing and growing quantum ecosystem that is producing functioning hardware, software, and algorithms today.

It's also important to mention that quantum devices and classical devices don't just solve problems differently; they solve different problems. And this will continue to be true in the years and even decades ahead. That's why it's not accurate to heap any quantum-related accolades on just quantum computers. The reality is that breakthroughs and disruptions will be the product of the combined power of classical and quantum computers working together in a hybrid architecture and "conducted" with workflow orchestration tools to solve problems.

The important takeaway here is that quantum's main role will be as a force multiplier to existing AI and machine learning solutions. For example, instead of just deep faking a celebrity video as you can do now, adding quantum to the mix could mean deep faking how lung cancer might spread in an individual based on their genetic profile and numerous other factors.

In fact, we have already entered a regime in which quantum devices, in certain, extreme edge case circumstances, can outperform their classical counterparts. This is referred to as *quantum supremacy*. These achievements are not yet scalable or practical for solving everyday useful problems, but future progress will eventually get us there, and the path to get there is already laid out before us.

In the meantime, there is a vibrant and growing ecosystem of government, academic, and business stakeholders—oftentimes working collaboratively—that is laying the groundwork for quantum's inevitable disruption of the status quo. Some of the industries that are expected to experience this disruption early include pharmaceuticals, healthcare, financial services, material science, logistics, cybersecurity, aerospace, and energy.

## It's All Connected

> "When we try to pick out anything by itself, we find it hitched to everything else in the universe."
>
> —John Muir

One of the features of the potential of quantum computing that is particularly enticing is how it has the potential to leverage the interconnectedness of so many variables and problems in consequential ways. Climate change, for example, does not take place in a vacuum. It directly impacts food security, which in turn impacts global trade, business (name any industry), economies, health, epidemiology, and social unrest, and this creates an added meta-dimensionality to how we compute predictive models. Quantum computing will give us the computing horsepower to address the dimensionality of this interconnectedness of myriad variables in intertwined complex systems.

While, as mentioned previously, quantum computing can—and will—be used by those with malicious intent, such as cybercriminals, the bulk of the intentional use cases, research, and work will go toward the betterment of society. Drug and vaccine development will speed up as molecules and biological systems are simulated in ways not possible today. Climate change models will be many levels beyond what we can generate today. Materials that don't exist now will be simulated and brought into existence. Investment portfolios will be optimized to an extent that all financial institutions dream of but currently cannot achieve. Hardware and software vendors will build and run their products in ways that are currently beyond the realm of imagination.

These are just a few examples of what we can expect to see in the years to come and are limited by what we can currently imagine and foresee, which represents just the tip of the possible outcome iceberg. We're unfortunately but inevitably bound to an orthodoxy that constrains our thinking. However, just as the advent of classical computers has fundamentally and forever changed our way of addressing problems, one consequence of using quantum computing will inevitably be to fundamentally require us to think about problems in a more probabilistic and scenario-driven way. This will be paradigm-changing for all of humanity on a foundational level—a truly seismic shift in capabilities and thinking. And this is one of the main motivations for founding a company to enable and participate in the creation of such a future.

## Thinking Differently

Back in the late 1990s, Apple ran a now-legendary ad campaign built around the message "Think Different." It featured that tagline supported by pictures and videos of transformational individuals such as Albert Einstein, Thomas Edison, and Martin Luther King, Jr. In a way, that is what quantum computing is soon going to help us do.

The "known unknown" with quantum computing is how it will help us look at things differently to come up with ideas—and the ability to process and compute them—that humans either couldn't conceive of or were too stuck in old, orthodox thinking to change. Think of the binary foundation of computing that consists of bits that can only be a one or a zero. Now, with the quantum model, computers can operate in a nonbinary mode because the quantum bits (called *qubits*) exist in what's called *superposition*, meaning they are in a state of limbo that is neither one nor zero, but rather an indeterminate, probabilistic mix of both "one-ness" and "zero-ness" until a quantum calculation is finally measured at the end.

Because of this quantum state, qubits can compute much more information and perform calculations exponentially faster than classical bits. It is not just better and bigger and faster. It is a fundamentally different way of processing the same information that changes the way we even formulate the problems that need to be solved. This forces the problem-solver to rethink the problem in new ways that add diversity in thinking at a fundamental level. Just as diversity of backgrounds and perspective confers advantages to human organizations, this diversity of thought in the problem-solving space will confer unique advantages in seeking outcomes and solutions to our most difficult and vexing problems.

This is the beautiful aspect of quantum computing: the as-yet-unknown possibilities that will manifest as negligently positive outcomes. The beautiful accidents. By definition, I cannot predict what they will be, nor can anyone else. But they are out there waiting to be born. I can't wait to see how this plays out!

# Appendix A: What Is Quantum Computing?

Philip L. Frana, *Associate Professor of Interdisciplinary Liberal Studies & Independent Scholars, James Madison University*

Quantum computing is a fundamentally unique way of processing information and calculating solutions to problems. Quantum computers are capable of operating in an extremely large number of states simultaneously, while classical computers can operate in only one state at any given moment. Because they work in a different way, many scientists believe that these quantum computers can deliver exponential speedups and solve problems that elude classical computers.

Classical computers ushered in the current Information Age, with all of its revolutionary digital advances: personal computing, Internet communication, smartphones, machine learning, and the knowledge economy generally. Classical computers encode and manipulate data in units known as *bits*. Today, these traditional general-purpose machines use billions of semiconductor parts known as *transistors* to switch or amplify electrical signals. A classical bit, like the power switch on your favorite electronic device, can be in one of two states at any given time, either 0 or 1. This is why classical information processing is said to be binary.

## How Quantum Computing Works

Quantum computers process information by exploiting the actions of subatomic particles, for example, electrons, ions, or photons. Quantum computers store information in quantum registers, which are in turn composed of quantum bits or qubits. These qubits are bounded only by the physical limits of superposition, entanglement, and interference. Superposition is a nonintuitive property of the subatomic world that permits qubits to exist in multiple states until some external measurement is taken. In the world of the quantum, for example, the state of an electron may be the superposition of the properties "spin up" and "spin down." A common analogy is Schrödinger's cat, which is both dead and alive until an observer peers inside the box. Qubits in superposition may be in a 0 state or a 1 state, but they may also be pointing in any other direction, which might be thought of in the quantum sense as some complex linear combination of 0 and 1. When the qubit is measured, the in-between "hidden" information collapses, and the new state will be binary, depending on whether the quantum state was closer to 0 or 1; if its amplitude is exactly in the middle, there is an equal probability of its resolving to either state. Upon measurement, the qubit becomes a classical bit.

Entanglement is another property of quantum physics that involves pairing and connection between particles in such a way that they cannot be described independently, even sometimes over great physical distances. Albert Einstein described entanglement as "spooky action at a distance." Bits in classical computers are independent from one another; a single bit does not exert any influence over any other. This is not true of quantum computers. In quantum computing, qubits can become entangled in such a way that they fall into a shared quantum state. Entangled qubits are no longer independent; manipulating one qubit can affect the probability distribution of the whole system. The number of states also becomes larger. One qubit is capable of holding two states at the same time (0 and 1). Two qubits can hold four states, three qubits give you eight states, four qubits sixteen states, and so forth. Sixty-four qubits yields 18,446,744,073,709,551,616 states, which a personal computer operating at a normal speed could cycle through in about 400 years. Each time a qubit is added, the number of simultaneous states doubles in a quantum computer, representing a huge advantage over a classical computer, which can be in only one state at a time. Theoretically, a quantum computer not affected by decoherence and noise (described in a moment) possesses truly massive processing power; 300 qubits could examine more possibilities than the number of atoms in the observable universe.

The final property of quantum mechanics that affects the operation of a quantum computer is interference. The mathematical description of qubits is represented in quantum mechanics by the wave function, a variable quantity that describes the isolated state of a quantum system of entangled qubits. When the wave functions of all of the entangled qubits are added together, we have both a description of the state of the quantum computer and also of interference. A common analogy here is the pattern of ripples of a body of water: sometimes the ripples join to make a bigger wave and sometimes when they come together produce stillness. Constructive interference increases the probability that the quantum computer's answer to a problem will be correct; destructive interference decreases that probability. Quantum algorithms are designed to choreograph this constructive and destructive interference and increase the probability that a qubit system collapses into useful measurement states. When contributions to the amplitude of entangled qubits reinforce one another, the probability of the right solution being recognized when the quantum computer's operator seeks a measurement is greatly increased. One of the tricks that can be carried out on a quantum computer is called *inversion about the mean*. One pass through the quantum circuit is unlikely to meaningfully increase the wave function value for the right answer. Too many iterations through the circuit can actually decrease the probability of rotating the initial state closer to the winner. In other words, an optimal mathematical floor and ceiling exists that increases the probability of identifying the correct item when the measurement is taken.

## Origins of Quantum Computing

The original idea for a quantum computer is ascribed to Soviet mathematician Yuri Manin who suggested the possibility in the introduction to his book *Computable and Uncomputable* in 1980. That same year, American physicist Paul Benioff, working at the French Centre de Physique Théorique, produced a paper in which he described a quantum mechanical model of a Turing machine. The very next year, Benioff and American theoretical physicist Richard Feynman delivered separate talks on quantum computing at the first Conference on the Physics of Computation at MIT. In his lecture "Simulating Physics with Computers," Feynman famously interjected a comment about how simulating a quantum system necessitates the construction of a quantum computer: "Nature isn't classical, dammit, and if you want to make a simulation of nature, you'd better make it quantum mechanical, and by golly it's a wonderful problem, because it doesn't look so easy."

Benioff and Feynman's papers fired the imaginations of scientists in the final decades of the 20th century. British theoretical physicist David Deutsch hoped that such a computer would make it possible to test the "many-worlds interpretation" of quantum physics, in which multiple universes are said to exist across space and time in parallel with our own universe. Deutsch advanced the idea of a quantum Turing machine (QTM), the first general and fully quantum model for computation in a 1985 paper published in the *Proceedings of the Royal Society*. By 1992, Deutsch and Australian mathematician Richard Jozsa found a computational problem that could be efficiently solved on a universal quantum computer with their Deutsch-Jozsa algorithm. The problem they identified cannot, it is thought, be solved efficiently on a classical computer. For this work, Deutsch is called the "father of quantum computing."

## Examples of Quantum Speedup: Shor's and Grover's Algorithms

Other researchers began developing their own quantum computer models and algorithms. One of the most famous is Shor's algorithm. In 1994, AT&T Bell Labs' applied mathematician Peter Shor unveiled a method for factoring large integers in polynomial time. For our purposes, polynomial-time algorithms can be thought of as "efficiently solvable" or "tractable." They are, as the NIST *Dictionary of Algorithms and Data Structures* defines them, "reasonable to compute".[1] Factorization consists of breaking down a number into smaller numbers that, when multiplied together, return the beginning number. It is trivially easy to multiply the small numbers (factors) together to produce the original number, and the traditional algorithm for doing so is fast, efficient, and known to every schoolchild. However, finding the original factors of numbers, and in particular very large numbers, is much more difficult. This is because the search space of possible factors is also very large.

Factoring and prime numbers are useful mathematical properties that are routinely used to secure communications on classical computers. Prime factorization—breaking down a number into the set of prime numbers that result in the original number when multiplied together—takes a very long time and in fact demands an algorithm that grows exponentially in running time as a function of the original number's size. In so-called public key cryptography, one person possesses a "public key," which is the product of two large primes. The public key is used to encrypt the message, and the two primes are used to decrypt the message. No published classical algorithm exists to find the prime factors of a public key in polynomial time; polynomial-time algorithms are informally described as "fast." Factoring a good public key is impractical because while it can be done, it takes too long using a classical computer searching for the original primes.

Theoretically speaking, Shor's algorithm can break existing encryption systems because factorization of primes can be achieved in polynomial time, which obviously excites tremendous interest among cryptographic specialists and anyone who wants to keep a vitally important system like email, an online bank account, or a nuclear weapons facility secure. Peter Shor's "fast" quantum algorithm attacks an exponential-time problem in mathematics, but it is too early to worry about a working quantum computer breaking current advanced encryption schemes. The key number size of public-key encryption systems like RSA continues to grow, which means that factoring time also increases. Also, while it might be possible to break encryption with something on the order of many thousands of qubits, noise and error-correcting codes will mean that significantly more qubits are needed. The current generation of universal quantum computers have no more than approximately 100 qubits; an encryption-breaking quantum computer would require a million qubits. Google Quantum AI recently announced at its annual developer conference that it intends to build a million-qubit machine by 2030, so perhaps we are living on borrowed time.

In 1996, another Bell Labs researcher, Lov Grover, presented the path-breaking paper "A Fast Quantum Mechanical Algorithm for Database Search" at the ACM Symposium on the Theory of Computing. This was followed by a more comprehensible piece in *Physical Review Letters* called "Quantum Mechanics Helps in Searching for a Needle in a Haystack." The advantage of Grover's quantum database search algorithm is that it provides a quadratic speedup* for one-way function problems usually accomplished by random or brute-force search; one-way function problems could involve searching for an item in an unsorted or unstructured list, optimizing a bus route, or solving a classic Sudoku puzzle. In other words, Grover's algorithm may be applied when a function is true for one input in the entire potential solution space, and false for all of the others. Rather than guessing one by one, which gives little information on what the right answer might be, Grover's algorithm leverages qubit superposition and interference to adjust the phases of various operations, increase the amplitude of the right item, and iteratively check and remove states that are not solutions. A measurement of the final state of the quantum computation returns the right item with certitude. Grover's algorithm works powerfully on computational problems where it is difficult to find a solution but relatively trivial to verify one.

## Policymaking and Partnerships

Excitement over these new discoveries and their potential for revolutionizing information processing gave rise to plans for

---

*Quadratic speedups are so-called second-degree polynomial time speedups. They involve the square exponent of a variable or unknown quantity and no higher power.

enhanced information sharing and policymaking and ultimately the prioritization and sequencing of national and international research efforts. The National Institute of Standards and Technology (NIST) and the Department of Defense (DoD) hosted the first U.S. government workshops on quantum computing in the mid-1990s. In 2000, theoretical physicist David DiVincenzo outlined the requirements necessary for constructing a quantum computer. These requirements are known as the DiVincenzo criteria and include such things as well-defined qubits, initialization to a pure state (complete knowledge of the system as opposed to indeterminacy or uncertainty), a universal set of quantum gates, qubit-specific measurement, and long coherence times. In 2002, an expert panel convened by Los Alamos National Laboratory released a Quantum Information Science and Technology Roadmap to capture the challenges involved in quantum computing, provide some direction on technical goals, and capture and characterize progress toward those goals through a variety of technologies and approaches. The panel decided to adopt the DiVincenzo criteria to evaluate the viability of various quantum computing approaches.

Evaluations of quantum computing models and approaches began yielding to instantiations in physical hardware and useful algorithms. In 1995, Christopher Monroe and David Wineland demonstrated the first quantum logic gate with trapped ions (the controlled-NOT)—an indispensable component for constructing gate-based quantum computers—publishing their results in *Physical Review Letters*. In 2005, researchers at the University of Michigan created a scalable and mass-producible semiconductor chip ion trap as a potential pathway to scalable quantum computing. In 2009, researchers at Yale University made the first solid-state, gate quantum processor. Two years later, D-Wave Systems of Burnaby, British Columbia, became the first company to market a commercial quantum computer. D-Wave's machine involves a unique approach to analog computing known as *quantum annealing*. Annealing processors are special-purpose technology; they are deployed against problems where the search space is discrete, with many local minima or plateaus, such as combinatorial optimization problems. It is not a universal quantum computer.

Yet, with the introduction of the original D-Wave, it became clear that fundamental advances in quantum hardware and software might yield extraordinary economic rewards and national security dividends. The research involved would be expensive and risky. A number of partnerships were forged in the early 2000s between private-sector companies and government agencies. Early buyers of D-Wave quantum computers included Google in alliance with NASA, Lockheed Martin Corporation in cooperation with the University of Southern California, and the U.S. Department of Energy's Los Alamos National Laboratory.

Google Research, NASA, and the Universities Space Research Association soon agreed that the value of quantum computers in solving intractable problems in computer science, and especially machine learning, was so great that they formally established a Quantum Artificial Intelligence Lab (QuAIL) at NASA's Ames Research Center in the Silicon Valley. NASA is interested in using hybrid quantum-classical technologies to attack some of the most difficult machine learning problems, such as generative unsupervised learning. IBM, Intel, and Rigetti are also chasing goals that would demonstrate quantum computational speedups over classical computers and algorithms in a variety of areas (sometimes termed *quantum supremacy* or *quantum advantage*). In 2017, University of Toronto assistant professor Peter Wittek founded the Quantum Stream in the Creative Destruction Lab (CDL). Despite Wittek's untimely death in a Himalayan avalanche, Quantum Stream continues to encourage scientists, entrepreneurs, and investors to pursue commercial opportunities in quantum computing and machine learning. Quantum Stream's technology partners include D-Wave Systems, IBM Q, Rigetti Computing, Xanadu, and Zapata Computing. Dozens of other startups and well-established companies are sprinting forward to create their own quantum computing technologies and applications, including the first quantum computing software company, 1QB Information Technologies (1QBit). In November 2021, IBM Quantum announced Eagle, a 127-qubit quantum processor. It is possible, however, that the leader in quantum computing is now the University of Science and Technology of China, which also in November 2021 claimed a 66-qubit superconducting quantum processor called Zuchongzhi and an even more powerful photonic quantum computer called Jiuzhang 2.0.

It is hard to know who has achieved primacy because verification and benchmarking of quantum computers remains a murky process and also because of the inherent diversity in current approaches and models of quantum computers. There is excitement surrounding a variety of models for manipulating a collection of qubits: gate model quantum computing, quantum annealing, adiabatic quantum computing (AQC), and topological quantum computing among them. There is also great diversity in methods for building physical implementations of quantum systems. Companies and research labs internationally are pursuing superconducting quantum computers, linear optical quantum computers, nitrogen-vacancy quantum computers, quantum computing with neutral atoms trapped in optical lattices, and a variety of other designs. More methods, approaches, and implementations may yet be undiscovered.

The physical implementation is important because quantum computers and qubits are devilishly difficult to control. Information stored in qubits can escape when the qubits become accidentally entangled with the outside environment, the measurement device and controls, or the material of the quantum computer itself. This seepage of quantum information is called *decoherence*. Qubits also need to be shielded physically from any kind of noise: changing magnetic and electrical fields, radiation from other electronic devices, cosmic rays from space, radiation from warm objects, and other rogue particles and waves. Making and manipulating high-quality qubits in quantum computers will require reducing decoherence and noise and also perhaps the invention of the sort of planning for fault-tolerance found in traditional computers. Quantum error correction is a multiply redundant scheme for spreading the information of one qubit and encoding it onto the highly entangled state of several other physical qubits. It is not known how many physical qubits will be needed to model a single logical qubit accessed by a quantum algorithm, but the number may be 100 to 10,000 times as high. Entangling, controlling, and measuring qubits have yet another major impediment familiar to generations of designers of classical computers: problems of scalability.

In 2018, President Donald Trump signed the National Quantum Initiative Act into law. The act is designed to plan, coordinate, and accelerate quantum research and development for economic and national security over a 10-year period. Funded under the National Quantum Initiative Act is the Quantum Economic Development Consortium™ (QED-C™), with NIST and SRI International as lead managers. Fundamental to the passage of the law is a shared recognition that quantum computing promises to contribute solutions to humanity's greatest and most difficult challenges in the areas of agriculture, biology, chemistry, climate and environment, communications, energy, healthcare, and materials science.

Quantum computer science is supported by a number of important online resources. The Quantum Algorithm Zoo, a comprehensive catalog of quantum algorithms, is managed by

Stephen Jordan in Microsoft Research's Quantum Systems group. IBM hosts the Quantum Experience, an online interface to the company's superconducting quantum systems and a repository of quantum information processing protocols. Qiskit is a software development kit (SDK) that has been open sourced for anyone interested in working with OpenQASM (a programming language for describing universal physical quantum circuits) and IBM Q quantum processors. Google AI in collaboration with the University of Waterloo, the "moonshot factory" X, and Volkswagen announced TensorFlow Quantum (TFQ) in 2020; TFQ is a Python-based open source library and framework for hands-on quantum machine learning.

## Quantum AI/ML

Quantum computing applications have already made headway in machine learning and AI, genomics and drug discovery, the chemical industry, molecular biology, cryptography, transportation and warehouse logistics, Internet communications, and simulation of quantum systems. Quantum simulation in particular could facilitate rapid prototyping of materials and designs, long before construction of parts or assemblies through CNC machining, injection molding, rapid tooling, or 3D printing. Currently, the top quantum computers are capable of simulating only a handful of particles and their interactions. But tantalizing clues are being found that may unravel the low-temperature behavior of exotic materials and superconductivity, help us understand the chemistry and production of environmentally friendly carbon-neutral fertilizers and cements, facilitate the design of next-gen EV batteries and solar panels, and model the complexities of flight mechanics, aerodynamics, and fluid dynamics in the aerospace industry.

Here are some of the more mind-blowing developments: Edward Snowden's 2014 leak of National Security Agency files confirmed the existence of the SIGINT initiatives "Penetrating Hard Targets" and "Owning the Net" to break any form of strong encryption, gain access to high-value secure digital communications networks, and design and attack Quantum Key Distribution (QKD) protocols. For these purposes, the agency planned to develop an $80 million quantum "god machine." In 2015, Unai Alvarez-Rodriguez of the University of the Basque Country in Spain shared research called "Artificial Life in Quantum Technologies," which he believes "paves the way for the realization of artificial life and embodied evolution with quantum technologies." In 2019, researchers at Ulm University in Germany observed evidence of quantum Darwinism in a test of synthetic diamond at room temperature. Quantum Darwinism is a theory that explains how our world of objective, classical physics emerges from the vagaries of the quantum world. Quantum Darwinism asserts that the "quantum-classical transition" is similar to the process of evolutionary natural selection. Physical properties selected from a bouillabaisse of possibilities become concrete because they are the "fittest" survivors. This is why, for instance, separate individuals can measure a quantum system and ultimately reach agreement on their findings. In just the last year, scientists have (a) announced a proof of concept for remote-sensing quantum radar, (b) created an unhackable integrated quantum communication network linking nodes over a total distance of 2,850 miles, and (c) developed a proposal to target and test potential quantum communications sent by extraterrestrials using existing telescope and receiver equipment.

The convergence of quantum computing and artificial intelligence (called quantum AL/ML [QAI]) will dramatically alter information science and technology, economic activity and social paradigms, regulatory frameworks, and political and security arrangements. The Fourth Industrial Revolution of GRIN technologies—genetics, robotics, information, and nanotechnologies—promises to soon give way to what the Japanese call "Society 5.0" and the Dutch term "Smart Humanity." The next revolutionary shift could produce a post-scarcity golden age where quantum AI for good holds sway and where advances in quantum technologies permit the universal democratization of access to limitless computational possibility. Or it could produce a postapocalyptic hellscape. Perhaps we are already living in that technological dystopia and should attempt to bolt for freedom.

Johannes Otterbach of the quantum-computer company Rigetti has remarked that quantum computing and machine learning are inherently probabilistic and thus natural bedfellows. Quantum computers could dramatically increase the speed of training in machine learning. Quantum machine learning will advance all three of the primary subcategories of ML: supervised learning, unsupervised learning, and reinforcement learning. Researchers are searching for quantum machine learning algorithms that demonstrate substantial speedups over classical algorithms and that overcome intractable exponential-time obstacles to problem solving and decision-making in the areas of sampling, search, optimization, pattern recognition, predictive- and risk-analytics, and simulation.

## Quantum AI/ML Applications

One powerful example of the intersection of quantum computing and AI is the development of working quantum algorithms for route and traffic optimization. Essentially, these quantum applications compute the quickest route for each individual vehicle in a fleet and optimizes it in real time. Toyota Tsusho

Corp—working in partnership with Microsoft and the quantum computing firm Jij—has demonstrated the potential of quantum routing algorithms to reduce the wait time at red lights by 20 percent. Volkswagen has also successfully tested quantum-computer enhanced navigation and route optimization applications on the CARRIS public bus fleet in Lisbon, Portugal, using a D-Wave machine. The goal of this test was to reduce traffic congestion and travel times during the Web-Summit technology conference. Volkswagen also developed a quantum simulation of the optimal routing of 10,000 taxis moving between the Beijing airport and the central business district 20 miles away using D-Wave technology. Such real-time quantum applications could also become useful in cross-functional supply-chain management and transportation logistics and in AI-powered autonomous cars and trucks.

Predictive and risk analytic QAI technology will aid in the forecasting, management, and disruption of hazards such as adverse geopolitical events or terror attacks, stock market crashes and financial panics, utility grid overloads, anthropogenic threats (climate change, habitat destruction, overexploitation of natural resources), social unrest, and future pandemics. Legal studies scholars are already examining the implications of a new field of "quantum jurisprudence" or "quantum AI law." Some scholars are even pondering the paradoxical implications and casuistry of criminality, where disputes, violations, and breaches of contract simply evaporate in the process of asking of questions about it. Total information awareness and quantum legal simulations and decision-making will make predictive pre-delinquency and policing more muscular, and precrime fighting (in the academic rather than the science fictional sense) more likely. On the other hand, quantum AI could also make destabilizing Cambridge Analytica–style political manipulations or Equifax-like data breaches more quotidian occurrences.

Artificial intelligence algorithms are also helping to decipher the physics of quantum systems. For example, leading-edge quantum sensing technology is used to detect extremely small variations in microgravity using solid state or photonic quantum computing systems. The technology is expected to advance the state-of-the-art in seismology, geological prospecting, electromagnetic field sensing, global positioning systems, measurement, microscopy, advanced radar, atomic clocks, magnetometers, and ultra-sensitive gravimeters. Quantum sensors could provide precise warnings of seismic events like earthquakes and volcanic eruptions, tsunamis, and silent-running enemy naval submarines.

Medical imaging technologies already involve serious use of computerized expert systems and complex pattern recognition software. There is a classic (manually applied) heuristic approach to melanoma diagnosis called ABCDE (asymmetry, border irregular, color distribution, diameter large, evolving mole). A convolutional neural network using machine learning has been trained on millions of images to apply ABCDE in the identification of skin lesions, melanomas, rashes, and other abnormalities. Another supervised learning algorithm called CheXNet outperforms expert radiologists at pneumonia screening and diagnosis. QAI in imaging, or *quantum radiomics*, promises to take these interpretive efforts to the next level. Quantum artificial neural networks may not spell the death knell of radiology—a long-overdue prediction—but could make the specialty more cost-effective and efficient by reading in microseconds the exponentially growing numbers of medical scans taken around the world—effectively pre-analyzing images, flagging ambiguous features, and helping humans avoid common errors attributed to boredom, inattention, and fatigue. Quantum radiomics could also attack the complex, real-time optimization problems of weighing and assessing the thousands of variables that contribute to the making of flexible and effective radiotherapy treatment plans for cancer patients.

## Quantum Ultra-intelligence

The promise and perils of quantum artificial intelligence are anticipated in an emerging literary subgenre called *quantum fiction*. Books that feature realistic or fanciful forms of QAI include *Factoring Humanity* (1998) and *Quantum Night* (2016) by Robert J. Sawyer, *Ghostwritten* (1999) by David Mitchell, *2312* (2012) by Kim Stanley Robinson, and *Antediluvian* (2019) by Wil McCarthy. The Japanese cyberpunk manga series *Battle Angel Alita: Last Order* (2000–2014) and *Kiddy Grade* (2002) are populated by several mysterious quantum AIs. Films include *Transformers* (2007) where a robot's "signal pattern is learning" using "quantum mechanics" and *Transcendence* (2014) with Johnny Depp. The Hulu TV miniseries *Devs* (2020) depicts a fictional quantum computing company and comments on the many-worlds interpretation of quantum mechanics and the effects of quantum technologies on determinism and free will. In the HBO television series *Westworld* (2016–present), a quantum artificial intelligence system named Rehoboam engineers and directs real-world society using its copious database. Many fictional quantum computers that have attained consciousness, in particular the laboring synthetic "geths" who are in conflict with their extraterrestrial humanoid masters the "quarians," populate the universe of the acclaimed military sf video game franchise *Mass Effect* (2007–present).

While these examples are all fiction, back here in reality some computer scientists have given their lives and careers over to engineering an artificial general intelligence (AGI) that possesses self-awareness, even to the point where it bootstraps itself to ultra-intelligence and unlocks the Technological Singularity. An emerging ultra-intelligence may enjoy a running start, as it will have instant access to the Penrose-Hameroff theory of quantum consciousness and neuro-inspired computer chips to create the blueprints for its self-designed quantum neural networks. It is unclear whether we will be able to maintain "human-in-the-loop" control over a self-aware QAI or cajole it into a superhero partnership with humanity (so-called collaborative QAI). Several experts responding to a 2021 Pew Research question on ethical AI doubted that a QAI would participate in curbing its own limits through something like a Quantum AI Constitutional Convention or Magna Carta for the Quantum Age.

Human beings are the only creatures on Earth with an almost unlimited capacity to learn, improve, and invent. Humans are very good at adding new complexity to their habitats. Indeed, the objective of so many who work in robotics, automation, and artificial intelligence today is not to restore human habitat to a blissful natural state but rather to create more and more of the fabricated world that we seemingly cannot do without. This impulse shapes the mutual goals of quantum computing and artificial intelligence and will have life-altering consequences. As the MIT physicist and ML specialist Max Tegmark has said, "Everything we love about civilization is a product of intelligence, so amplifying our human intelligence with artificial intelligence has the potential of helping civilization flourish like never before—as long as we manage to keep the technology beneficial."

# Appendix B: What Is Artificial Intelligence?

Philip L. Frana, *Associate Professor of Interdisciplinary Liberal Studies & Independent Scholars, James Madison University*

Artificial intelligence (AI) is a rapidly evolving field rooted in computing and the cognitive sciences. As an academic field, AI involves research on intelligent agents that perceive and respond to environments like cyberspace or the physical world to achieve specific goals. Siri and Alexa chatbots are examples of intelligent agents, as are the sensor and actuator-based systems implanted in Roomba vacuum cleaners and Tesla cars. More generally, AI encompasses both real and fictional efforts to imitate human (and animal) intelligence and creativity with machines and code.

Human behaviors and characteristics of interest to AI researchers may involve pattern recognition, problem-solving and decision-making, learning and knowledge representation, communication, and emotions. Some advances in AI are stunning enough to garner millions of views on social media, but for every Boston Dynamics robot performing synchronized gymnastics or Disney Stuntronic Spider-Man doing spectacular acrobatic tricks, there are dozens of AI applications (recommendation and search engines, banking and investment software, shopping and pricing bots) that are so commonplace that we hardly remark upon their near-magical effectiveness anymore. In the popular imagination, "AI is whatever hasn't been done yet;" AI for most people is digital pixie dust.

Dreams of thinking machines are as old as civilization. Hesiod in c.700 BC tells the tale of the lethal autonomous robot Talos, who protected Crete by tossing giant rocks at enemy ships. Three centuries later a group of spirit movement machines (*bhuta vahana yanta*) were according to legend forged to protect the relics of the Buddha. In medieval times, Roger Bacon purportedly created a talking bronze head that, like Siri or Alexa, could answer queries. One of the earliest English-language accounts of machines with human-like intelligence is Samuel Butler's 1872 utopian novel *Erewhon*. Butler's machines are conscious and able to self-replicate. In the 20th century, fictional depictions of artificial intelligence found homes in the stories of Isaac Asimov, Philip K. Dick, and William Gibson. Hollywood is also enamored of sentient computers, producing classic films such as *2001: A Space Odyssey* (1968), *Blade Runner* (1982), and *The Terminator* (1984), among many others. Themes that are abundant in fiction about AI include authenticity, personhood, companionship, loneliness, dystopia, and immortality.

The origins of artificial intelligence as actual science are interdisciplinary. One source of artificial intelligence ideas is cybernetics, which sought to understand the role of mammalian neural pathways and connections that produce homeostasis and intelligent control. In the 1940s, the Teleological Society and Macy Conferences formed to tackle problems important to understanding human physiology, creating servomechanisms for use in factories and weapon systems, and envisioning super intelligent "giant brains." These organizations incubated the ideas of several pioneers important to the development of AI, including

John von Neumann, Warren McCulloch, Walter Pitts, and Claude Shannon. Cybernetic and connectionist models that showed how biological organisms self-regulate, interact with the environment, and achieve goals continue to inspire pathbreaking efforts in system theory, artificial neural networks, and artificial intelligence.

A second wellspring of ideas about AI derives from cognitive psychology. Experimental psychologists seeking to move away from behaviorism invented the computational theory of mind in the 1950s, and this movement is now called the Cognitive Revolution. The computational theory of mind combines the information theory work of Claude Shannon; Alan Turing's conception of mental activity as computation; Allen Newell and Herbert Simon's information processing models of human perception, memory, communication, and problem solving; and Noam Chomsky's generative linguistics. Cognitive psychology tackles a number of problems in human and artificial intelligence, including recognition, attention, memory, and psycholinguistics.

A third source for AI is rule-based and symbolic representations of problems, also known as *good old-fashioned AI* (GOFAI). In addition to Newell and Simon, other active proponents of this approach include Marvin Minsky, John McCarthy (who coined the term *artificial intelligence*), and Edward Feigenbaum. GOFAI in the latter half of the 20$^{th}$ century nurtured a broad range of knowledge-based expert systems that emulated human decision-making. AI systems were created for several academic fields and commercial applications. DENDRAL was designed to detect and identify complex organic molecules, potentially useful on automated NASA planetary missions. MYCIN diagnosed and recommended therapies for blood infections. INTERNIST-I encoded the expertise of a doctor of internal medicine. The Cyc project to create an expert system for "common sense" has spanned almost four decades. While a few AI developers still assert that an expert system might eventually approach the versatility of a human thinker, most now think that artificial neural networks and deep learning, or some combination of neural and symbolic approaches, have the greatest potential to approach the artificial general intelligence (AGI) found in speculative fiction.

Machine learning is often referred to as *artificial intelligence* but is actually a particularly productive subfield focused on using computer algorithms to build systems that autonomously learn from a given database and/or experiences. Machine learning systems learn gradually, much in the way humans are thought to learn. The goal, notes AI pioneer Arthur Samuel, is to implant in machines "the ability to learn without explicitly being programmed." Computer scientist Pedro Domingos defines "five tribes" within the subfield of machine learning: symbolists (inspired by logic and induction), connectionists (inspired by neural networks), evolutionaries (genetic development and transformation), Bayesians (statistics and probability), and analogizers (psychology and optimization). Machine learning platforms have improved medical care, facial and speech recognition, predictive analytics, warehouse management and transportation logistics, and many other workflows and tasks.

Work in machine learning is today divided into three broad types: supervised, unsupervised, and reinforcement. Supervised learning algorithms depend on labeled training data provided by human specialists. Here the machine learning model trains on input examples to classify, assess, or make predictions about similar new data. An example is using samples of spam email to design a spam filtering system. Unsupervised machine learning algorithms search for interesting patterns or structure in unlabeled datasets. The objective here is difficult to achieve, as the model is asked to provide valuable insights without training. An example might involve detecting and differentiating between groups of customers that have not otherwise been identified. Reinforcement learning depends on intelligent agents that react directly with the environment to achieve rewards or attain goals by trial, error, and feedback. Reinforcement learning is widely used to help train AIs to play games and drive automobiles.

Deep learning is a subfield of machine learning inspired by the structures and functions of the human brain. Resurgent interest in multilayer artificial neural networks (ANNs) and deep learning is producing exciting advances in speech recognition and natural language processing (NLP), computer vision and image recognition, neuromorphic computing, sustainability science, bioinformatics, and smart devices and vehicles. Deep learning powers the top machine translation engines (SYSTRAN, Google Translate, Microsoft Translator), imaging technologies (DeepFace, CheXNet, StyleGAN), environmental monitoring systems (Green Horizons, Wildbook, PAWS) and computational creativity applications (Deep Dream, MuseNet, WaveNet).

Today, much interdisciplinary work in artificial intelligence occurs in university and corporate computer science laboratories and within the fuzzy boundaries of cognitive science. Cognitive science is a multidisciplinary venture propelled by researchers in artificial intelligence, android science, biological information processing, computational neuroscience, cognitive psychology, human and animal cognition, linguistics and anthropology, neurology, and philosophy of consciousness. Notable 21st century computer scientists straddling multiple disciplines in cognitive science are Demis Hassabis (DeepMind), Geoffrey Hinton (Google Brain), and Fei-Fei Li (Stanford HAI).

Artificial intelligence research is grounded, and in many ways held accountable, by the hard questions of ethics and consciousness in philosophy. The ethics of AI extend back to the Three Laws of Robotics offered up in Isaac Asimov's short story "Runaround" (1942). The three laws still attract conversation but have largely been supplanted by other issues, especially the "black box"* of AI decision-making and questions of machine autonomy and human complacency. Instances of algorithmic bias and discrimination are common and growing. Problems of algorithmic accountability and governance are partially addressed in the European Union's General Data Protection Regulation (GDPR) and "Ethics Guidelines for Trustworthy AI" (2018), and in the proposed Artificial Intelligence Act (2021). Elsewhere, governments and corporations are creating directorates to recommend adoption of new policy frameworks, with varying levels of success. Explainable AI (XAI) refers to multiple approaches and design choices that reduce the potential for bias while making the inner workings of AI models transparent to human observers. Prominent organizations advocating for equitable and accountable artificial intelligence include the Algorithmic Justice League, the Partnership on AI (Amazon, Facebook, IBM, Google, and Microsoft), and the Global Partnership on AI. The goal of AI for Good's global summits is to identify artificial intelligence solutions that accelerate progress toward the United Nations Sustainable Development Goals.

Other professional organizations are also looking closely at the moral conduct of machines and their designers. AI autonomy in motor vehicles, autonomous weapons systems, and caregiver robots opens up a host of new opportunities and threats. The Society of Automotive Engineers International (SAE) defines six levels of driver automation. At level 0, the human driver is in full

*Black-box AI is any artificial intelligence system that cannot produce accountable and transparent results that include an explanation about how the results were obtained. Black-box AI makes biased data, unsuitable modeling techniques, and incorrect decision-making more difficult if not impossible to detect.

control of all responses to the environment and emergent threats. At level 1 a human driver is assisted by an automated system for longitudinal or latitudinal (lane centering) control. At level 2 the automated system provides steering, braking, and acceleration support (adaptive cruise control). At level 3 the human is not driving until the automated system requests that the human retake control. An example of level 3 support is "traffic jam chauffer." At level 4 the automated system no longer requires the human ever assume control but works only under specific conditions. An example here is a "local driverless taxi." At level 5 the vehicle is capable of driving itself under all conditions and without a human being present. Level 2 is the highest level of autonomy available with General Motor's Super Cruise, Nissan's ProPilot, or Tesla's Autopilot. As of 2021, no cars have yet reached the level 3, 4, or 5 stage of autonomy.

Lethal autonomous weapons systems (LAWS) are similarly divided into levels of AI autonomy. Human-in-the-loop weapons choose their targets and destroy them only under direct human authority. Human-on-the-loop weapons are monitored but largely free to deliver force autonomously. Overriding a primary directive in human-on-the-loop systems may require the hair-trigger response of a human supervisor. Human-out-of-the-loop weapon systems identify, target, and destroy enemies without any human oversight. Examples of powerful AI weapons are the U.S. Navy's MK 15 Phalanx CIWS ("sea wiz") and the Israeli IAI Harpy "suicide" drone. The Harpy is categorized as a loitering munition that autonomously flies over an area until it finds a target to attack; the Harpy has aroused concern that it violates the laws of war.

More constructively, caregiver robots provide aid as assistants and companions to vulnerable populations, such as children, the disabled, the mentally ill, and the elderly. AI caregiver technology is available or being tested in many countries but is most common in Japan where cultural acceptance and an aging population have stimulated sales of plush robot baby beluga whales (Paro), robotic therapy dogs (AIBO), and autonomous humanoid patient-lifting robots (Robear).

Roboethicists are engaged in understanding the moral conduct of human creators of artificially intelligent robots. The Foundation for Responsible Robotics and the European Robotics Research Network (EURON) recognize the importance of human accountability in the development of AI systems. Other experts are thinking about full-fledged robot ethics and moral machines; one goal is implanting artificial ethical capability into every autonomous machine.

AI will undoubtedly adversely affect the nature and future of work around the world. AI threatens to throw millions of retail sales workers and managers, accountants and bookkeepers, factory workers, and journalists out of work. Oppositely, some experts in the trucking industry predict that a persistent driver shortage will trigger a full-scale switch to autonomy before 2030; there are today more than 3.5 million truck drivers in the United States alone.

The impact on life and work will be even greater if advances are made in quantum artificial intelligence (QAI) and superintelligence. Linking quantum processors to AI could make possible the autonomous management of traffic in an entire city, pharmaceutical discovery much less costly and arduous, or render digital encryption of military secrets obsolete. Respected authorities such as Nick Bostrom, Ray Kurzweil, and Murray Shanahan warn that a Technological Singularity facilitated by ultraintelligent AIs could wreak havoc on human civilization, perhaps even precipitating an extinction event. On the other hand, an exponentially growing artificial intelligence could just as easily bring about an end to catastrophic climate change, overpopulation, or cycles of intergenerational poverty.

# Glossary

**Artificial intelligence (AI):** See Appendix B for definition and discussion.

**Bias**: A problem that happens when an algorithm produces prejudiced results due to faulty assumptions in the machine learning process.

**Complex systems**: A complex system is a system composed of many components and subcomponents that interact with each other and whose behavior is intrinsically difficult to model due to the dependencies, competitions, relationships, or great numbers of interactions between their parts or between the system and its surroundings. Examples of complex systems include the human brain, biological organisms, global climate, and infrastructure, such as the power grid, transportation or communication systems, complex software and electronic systems, and social and economic organizations.

**Complexity theory**: Complexity theory is the study of complex systems. While it is a relatively new field of study, it covers a wide range of disciplines in the physical, biological, and social sciences.

**Decoherence**: Quantum decoherence is the loss of quantum coherence. As long as there exists a definite phase relation between different states, the system is said to be coherent.

**Edge case**: In software engineering, an edge case is a problem or situation caused by a parameter exceeding the bounds the system was designed to accept. In other words, edge cases occur only at extreme operating parameters.

**Encryption**: Encryption transforms data to lock information using an algorithm. A password or "key" is used to unlock the data and converts the information to make the original information readable.

**Entanglement**: This is a phenomenon that occurs when a group of particles are generated, interact, or share spatial proximity in a way such that the quantum state of each particle of the group cannot be described independently of the state of the others, even when the particles are separated by a large distance.

**Error rate**: Current quantum computers typically have error rates near one in a thousand ($10^3$), but many practical applications call for error rates as low as one in a quadrillion ($10^{15}$). See also **fault tolerance**.

**Fault tolerance**: The nature of quantum computers means that they will not be able to perform gate operations perfectly—some error is unavoidable. The fault tolerance of a quantum computer reflects its ability to protect quantum information from such errors (due to decoherence and other quantum noise). However, although Noisy Intermediate-Scale Quantum (NISQ) computers are realizable in the near-term, fully fault-tolerant quantum computing is not likely to happen for some time because of the large number of physical qubits needed.

**Game theory**: This is the study of strategies, examining what is the best choice given multiple (or even infinite) choices in one or multiple interactions with one or more players.

**Grover's algorithm**: Grover's algorithm is a quantum algorithm used for searching an unsorted database. It was invented by Lov Grover in 1996.

**Hilbert space**: Hilbert space, in mathematics, allows generalizing the methods of linear algebra and calculus from the two-dimensional and three-dimensional Euclidean spaces to spaces that may have an infinite dimension.

**Indeterminacy**: A principle in quantum mechanics stating that it is impossible to accurately measure both the position and the momentum of very small particles at the same time.

**Machine learning**: See Appendix B for definition and discussion.

**Monte Carlo simulation**: Monte Carlo simulations are algorithms that use repeated random sampling to obtain numerical results. The main concept is to use randomness to solve problems that might not be random or deterministic.

**Moore's law**: This is a technology trend first observed by Gordon Moore, who noticed that transistor-based computers appear to double processor speed roughly every two years.

**Neural network**: A neural network (also known as an artificial neural network [ANN] or simulated neural network [SNN]) is a series of algorithms that endeavors to recognize underlying relationships in a dataset through a process that imitates how the human brain functions.

**Noise**: Quantum noise refers to the fluctuations of signal, that is, noise arising from quantum fluctuations.

**Noisy intermediate-scale quantum (NISQ)**: A term first used by John Preskill in 2018, noisy intermediate-scale quantum processors contain about 50 to a few hundred qubits. These processors are not sophisticated enough to achieve robust **fault tolerance**.

**Quantum advantage**: Quantum advantage refers to the achievement of processing a real-world problem faster on a quantum computer than on a classical computer. This is sometimes also called **quantum supremacy**.

**Quantum bit**: A quantum bit is the basic unit of information in quantum computing, the quantum equivalent of a classical binary bit. Just like classical bits, a quantum bit must have two states: 0 and 1. Unlike a classical bit, a quantum bit can also exist in **superposition** states, be subjected to incompatible measurements, and even be entangled with other quantum bits. Being able to use superposition, **quantum interference**, and **entanglement** makes qubits very different and much more powerful than classical bits. There are several kinds of qubits, including spin, trapped atoms and ions, photons, and superconducting circuits. Physical qubits in a computer refer to the number of qubits in the quantum computer. Logical qubits are groups of physical qubits used as a single qubit in processor operations.

**Quantum computer**: See Appendix A for definition and discussion.

**Quantum interference**: Quantum interference states that particles not only can be in more than one place at the same time (through **superposition**) but that a single particle, i.e., a photon (light particles), can cross its own trajectory and interfere with the direction of its own path. In other words, the wave function interferes with itself.

**Quantum supremacy**: See **quantum advantage**.

**Qubit**: See **quantum bit**.

**Shor's Algorithm**: In 1995, Peter Shor proposed a polynomial-time quantum algorithm for factoring a useful, real-life problem. Shor's algorithm was the first nontrivial quantum algorithm showing a potential of "exponential" speedup over classical algorithms.

**Superposition**: Superposition is the ability of a quantum system to be in several states at the same time until it is observed or measured.

**Turing test**: The Turing test was originally conceived by Alan Turing in 1950. The test evaluates a machine's ability to demonstrate intelligent behavior indistinguishable from a real person. If an evaluator cannot tell the difference between the machine and a real person, the machine is said to have passed the Turing test.

# References

## Foreword


[1] Kurzweil, R. (2006) *The Singularity is Near: When Humans Transcend Biology*. Penguin Books.

[2] Bostrom, N. (2014) *Superintelligence: Paths, Dangers, Strategies*. Oxford University Press.

[3] Asimov, I. (1942) "Runaround," *Astounding Science Fiction*.

[4] Metz, C. (2015) "Elon Musk's Billion-Dollar AI Plan is About Far More Than Saving the World," *Wired*, www.wired.com/2015/12/elon-musks-billion-dollar-ai-plan-is-about-far-more-than-saving-the-world (Accessed: 04 July 2016).

[5] Brin, D. (1998) *The Transparent Society, Will Technology Force us to Choose Between Privacy and Freedom?* Perseus Books.

[6] Hutter, M. (2004) *Universal Artificial Intelligence, Sequential Decisions Based on Algorithmic Probability*. Springer.

[7] Russell, S. & Norvig, P. (2009) *Artificial Intelligence: A Modern Approach*, 3rd ed. Pearson.

[8] www.deepmind.com (Accessed: 04 July 2016).

[9] Lewis, M. (2014) *Flash Boys: A Wall Street Revolt*. W.W. Norton & Company.

[10] www.ibm.com/watson (Accessed: 29 June 2016).

[11] OpenWorm, Artificial Brains, The Quest to Build Sentient Machines, www.artificialbrains.com/openworm (Accessed: 04 July 2016).

[12] Hanson, R. (2016) *The Age of Em: Work, Love, and Life When Robots Rule the Earth*. Oxford University Press.

[13] Brin, D. (1980–1999) Various publications under "Uplift Universe," www.davidbrin.com/uplift.html (Accessed: 04 July 2016).

[14] Brin, D. (2012) *Existence*. Orbit.

[15] Kelly, K. (2016) *The Inevitable: Understanding The 12 Technological Forces That Will Shape Our Future*. Viking.

[16] Barrat, J. (2015) *Our Final Invention: Artificial Intelligence and the End of the Human Era*. St. Martin's Griffin.

[17] Storrs Hall, J. (2007) *Beyond AI: Creating the Conscience of the Machine*. Prometheus Books.

[18] Leonhard, G. (2015) *Humanity Vs Technology A Short Film by Gerd Leonhard*, youtu.be/DL99deFJYaI (Accessed: 04 July 2016).

[19] Alang, N. (2016) "Life in the Age of Algorithms: As Society Becomes More Wedded to Technology, It's Important to Consider the Formulas that Govern our Data," *New Republic*, newrepublic.com/article/133472/lifeagealgorithms (Accessed: 04 July 2016).

[20] Havens, J. (2016) *Heartificial Intelligence: Embracing Our Humanity to Maximize Machines*. TarcherPerigee.

[21] Kurzeil, R. (2000) *The Age of Spiritual Machines: When Computers Exceed Human Intelligence*. Penguin Books.

[22] Markoff, J. (2015) Machines of Loving Grace: The Quest for Common Ground Between Humans and Robots, Ecco.

[23] Brin, D. (1999) *Foundation's Triumph*. Orbit.

[24] Byford, S. (2016) "Sony is Working on a Robot That Can 'Form an Emotional Bond' with People," *The Verge*, www.theverge.com/2016/6/29/12057408/sony-robot-emotion-vr-sensors (Accessed: 04 July 2016).

[25] Robotic Intelligence, "Sony To Create Robot That Can Form an Emotional Bond with People," *Futurism*, 30 June 2016, futurism.com/sony-to-create-robot-that-can-form-emotional-bond-with-people (Accessed: 04 July 2016).

[26] futureoflife.org (Accessed: 04 July 2016).

[27] Welsh, J. (2015) "Researchers Say this is the most Impressive Act of Artificial Intelligence they've Ever Seen," *Business Insider*, www.businessinsider.com/artificial-intelligence-playing-video-games-2015-11

[28] Allen, C. (2011) "The Future of Moral Machines," The Opinion Pages, *The New York Times*.

[29] Nadella, S. (2016) "The Partnership of the Future: Microsoft's CEO Explores How Humans and A.I can work Together to Solve Society's Greatest Challenges," www.slate.com/articles/technology/future_tense/2016/06/microsoft_ceo_satya_nadella_humans_and_a_i_can_work_together_to_solve_society.html (Accessed: 04 July 2016).

[30] Hanson, R. (2016) "Future Fears: Overcoming Bias," www.overcomingbias.com/2016/06/future-fears.html (Accessed: 04 July 2016).

[31] Hanson, R. & Yudkowsky, E. (2013) *The Hanson-Yudkowsky AI-Foom Debate*. Machine Intelligence Research Institute.

[32] Committee on Legal Affairs. (2016) "With Recommendations to the Commission on Civil Law Rules on Robotics" (draft report), European Parliament 20142019.

## Chapter 2

[1] Dunbar, R. (1998) *Grooming, gossip, and the evolution of language*. Harvard University Press.

## Chapter 3

[1] Furman III, D. & Musgrave, P. (2017) "Synthetic Experiences: How Popular Culture Matters for Images of International Relations," *International Studies Quarterly* 61 (3): 503–516, doi.org/10.1093/isq/sqx053.

## Chapter 5

[1] Fowler, M., et al. (2018) "Surface codes: toward practical large-scale quantum computation," *Physical Review A*, 86, 032324.

## Chapter 6

[1] Thompson, S. & Parthasarathy, S. (2006) "Moore's law: the future of Si microelectronics," *Materials today*, 9(6):20–25.

[2] Turing, A. (2009) "Computing machinery and intelligence," In *Parsing the turing test*, pages 23–65. Springer.

[3] Thompson, S. & Parthasarathy, S. (2006) "Moore's law: the future of Si microelectronics," *Materials today*, 9(6):20–25.

[4] Hamilton, K. (2020) "Accelerating scientific computing in the post-moore's era," *ACM Transactions on Parallel Computing (TOPC)*, 7(1):1–31.

[5] Feynman, R. (1982) Simulating physics with computers. *Int. J. Theor. Phys*, 21(6/7).

[6] Preskill, J. (2018) Quantum computing in the nisq era and beyond. *Quantum*, 2:79.

[7] Farhi, E., et al. (2014) A quantum approximate optimization algorithm. *arXiv preprint arXiv:1411.4028.*

[8] Harrigan, M., et al. (2021) "Quantum approximate optimization of non-planar graph problems on a planar superconducting processor," *Nature Physics*, pages 1–5.

[9] Preskill, J. (2018) Quantum computing in the nisq era and beyond. *Quantum*, 2:79.

[10] Huang, H., et al. (2021) "Information-theoretic bounds on quantum advantage in machine learning," *arXiv preprint arXiv:2101.02464.*

## Chapter 7

[1] Gidney, C. & Ekerå, M. (2019) "How to factor 2048 bit RSA integers in 8 hours using 20 million noisy qubits," `arxiv.org/pdf/1905.09749.pdf` (Accessed: November 30, 2020).

[2] Dridi, R. & Alghassi, H. (2016) "Prime factorization using quantum annealing and computational algebraic geometry," `arxiv.org/abs/1604.05796` (Accessed: 30 November 2020).

[3] Wang, B. (2020) "Prime factorization algorithm based on parameter optimization of Ising model," *Scientific Reports* 10 (7106).

[4] Peng, W., et al. (2019) "Factoring larger integers with fewer qubits via quantum annealing with optimized parameters," *Science China Physics, Mechanics & Astronomy* 62 (60311).

[5] Eisenbach, T., et al. (2020) *Cyber Risk and the U.S. Financial System: A Pre-Mortem Analysis*. Federal Reserve Bank of New York.

[6] Boston Consulting Group (2019). Global Wealth 2019: Reigniting Radical Growth.

[7] IBM X-force Incident Response and Intelligence Services (2020) "X-Force Threat Intelligence Index."

## Chapter 9

[1] Solon, O. (2017) "Alibaba Founder Jack Ma: AI Will Cause People 'More Pain than Happiness,'" *The Guardian*, `www.theguardian.com/technology/2017/apr/24/alibaba-jack-ma-artificial-intelligence-more-pain-than-happiness`.

[2] Kühn, M., et al. (2019) "Accuracy and Resource Estimations for Quantum Chemistry on a Near-term Quantum Computer," *Journal of Chemical Theory and Computation* 15, no. 9: 4764–4780, doi.org/10.1021/acs.jctc.9b00236.

[3] Kawashima, Y., et al. (2019) "Optimizing Electronic Structure Simulations on a Trapped-ion Quantum Computer using Problem Decomposition," *Communications Physics* 4, no. 245 (2021): 1–9. doi.org/10.1038/s42005-021-00751-9. Lichfield, G. (2019) "Google CEO Sundar Pichai on Achieving Quantum Supremacy," *MIT Technology Review*,www.technologyreview.com/2019/10/23/102523/google-ceo-quantum-supremacy-interview-with-sundar-pichai.

[4] World Health Organization. (2018) *Global Status Report on Road Safety*, www.who.int/publications/i/item/9789241565684.

[5] U.S. Energy Information Administration. (2022) *Short-Term Energy Outlook*, www.eia.gov/outlooks/steo/report/global_oil.php.

[6] Brown, A., et al. (2015) "An Analysis of Possible Energy Impacts of Automated Vehicles," in *Road Vehicle Automation*, edited by Gereon Meyer and Sven Beiker, pp. 137–153. Cham, Switzerland: Springer.

[7] TotalEnergies. (2018) "Total to Develop A.I. Solutions with Google Cloud,"corporate.totalenergies.us/home/media/list-news/total-develop-ai-solutions-google-cloud.

[8] PricewaterhouseCoopers and Microsoft, *How AI Can Enable a Sustainable Future*, p. 8, www.pwc.co.uk/sustainability-climate-change/assets/pdf/how-ai-can-enable-a-sustainable-future.pdf.

[9] Strubell, E., et al. (2019) "Energy and Policy Considerations for Deep Learning in NLP," in *Proceedings of the 57th Annual Meeting of the Association for Computational Linguistics*, pp. 3645–3650. Florence, Italy: Association for Computational Linguistics, arxiv.org/pdf/1906.02243.pdf.

[10] Bucher, T. (2017) "The Algorithmic Imaginary: Exploring the Ordinary Affects of Facebook Algorithms," *Information, Communication & Society* 20, no. 1: 30–44, doi.org/10.1080/1369118X.2016.1154086.

[11] Faggella, D. (2019) "Does the Environment Matter After the Singularity?" danfaggella.com/environment.

[12] Lehman, J. & Miikkulainen, R. (2015) "Extinction Events Can Accelerate Evolution," *PLoS ONE* 10, no. 8: e0132886, doi.org/10.1371/journal.pone.0132886.

[13] Yudkowsky, E. (2008) "Artificial Intelligence as a Positive and Negative Factor in Global Risk," Machine Intelligence Research Institute, p. 27, intelligence.org/files/AIPosNegFactor.pdf.

# Chapter 12


[1] Hoffmann, R. & Malrieu, J.-P. (2020) "Simulation vs Understanding: A Tension, in Quantum Chemistry and Beyond. Part A. Stage Setting," *Angewandte Chemie*, 59, 12590-125610.

[2] Hoffmann, R. & Malrieu, J.-P. (2020) "The March of Simulation, for Better or Worse," *Angewandte Chemie*, 59, 13156-13178.

[3] Hoffmann, R. & Malrieu, J.-P. (2020) "Toward Consilience," *Angewandte Chemie*. 59, 13694-13710.

[4] Kuang, C. (2017) "Can A.I. Be Taught to Explain Itself?" *New York Times Sunday Magazine*, p. MM46.

[5] Mozur, P. (2019) Facial Scans Tighten China's Grip on a Minority," *The New York Times*, p. 1, A8.

[6] Etzioni, A. & Etzioni, O. (2017) "The Pros and Cons of Autonomous Weapons Systems," *Military Review*, May-June 2017, 72–80.

[7] Thom, R. & Noël, É. (2009) "Prédire n'est pas expliquer," Flammarion.

[8] Dirac, P. A. M. (1929) *Proceedings of the Royal Society of London*. Series A, Vol. 123, 714–733.

[9] Levi, P. (1975) *The Periodic Table*.

[10] Sacks, O. *Uncle Tungsten: Memories of a Chemical Boyhood*.

[11] Gray, T. (2009) *The elements: a visual exploration of every known atom in the universe*. New York: Black Dog & Leventhal Publishers.

[12] Gray, T. (2018) *Molecules: the elements and the architecture of everything*. New York: Black Dog & Leventhal Publishers.

[13] Liang, Y. www.beautifulchemistry.net; www.12molecule.com/about; www.envisioningchemistry.com.


# Chapter 14


[1] Theis, T. N. & Wong, H. S. P. (2017) "The end of moore's law: A new beginning for information technology," *Computing in Science & Engineering*, 19(2), 41–50.

[2] Sevilla, J. & Moreno, P. (2019) "Implications of Quantum Computing for Artificial Intelligence alignment research," *arXiv preprint arXiv: 1908.07613*.

[3] Karlan, B. & Allen, C. (in prep.) "Engineered wisdom for learning machines."

[4] Kearns, M. & Roth, A. (2019) *The ethical algorithm: The science of socially aware algorithm design*. Oxford University Press.

[5] Bainbridge, L. (1983). "Ironies of automation," In *Analysis, design and evaluation of man–machine systems* (pp. 129–135). Pergamon.

[6] Lieder, F. & Griffiths, T. L. (2020). "Resource-rational analysis: understanding human cognition as the optimal use of limited computational resources," *Behavioral and Brain Sciences*, 43.

[7] Hagar, A. (2003). "A philosopher looks at quantum information theory," *Philosophy of Science*, 70(4), 752–775.

## Chapter 15

[1] Hillis, W. D. (1998) *The pattern on the stone: The simple ideas that make computers work*. New York: Basic Books.

[2] Grover, L. K. (1996) A fast quantum mechanical algorithm for database search, *Proceedings, 28th Annual ACM Symposium on the Theory of Computing*, p. 212.

[3] Clarke, A. (1982) *2010: Odyssey Two*, 1st ed. Ballantine Books.

## Chapter 16

[1] Kurzweil, R. (2012) *How to Create a Mind*. New York: Viking Press.

[2] AUTONOMOUS WEAPONS: AN OPEN LETTER FROM AI & ROBOTICS RESEARCHERS, 2016, futureoflife.org/open-letter-autonomous-weapons/?cn-reloaded=1.

[3] Pelton, J. (2010) *Megacrunch: Ten Survival Strategies for 21st Century Challenges*. London: PM Associates.

[4] Partners, R. (2018) "High Tech and High Touch-What does it take to survive in the emerging economy?" www.ryanpartners.net/blog/november_2011/high_tech_and_high_touch-what_does_it_take_to_surv.

[5] Thompson, C. (2018) "Elon Musk warns that creation of a 'god-like' AI could doom mankind to an eternity of robot dictatorship," *Business Insider*, www.businessinsider.com/elon-musk-says-ai-could-lead-to-robot-dictator-2018-4.

[6] Wahid, S. "The Imaginal Within The Cosmos: The Noosphere and Artificial Intelligence," www.academia.edu/38589357/The_Imaginal_Realm.

## Chapter 17

[1] mywaverly.com.

## Chapter 18

[1] Aaronson, S., Atia, Y. & Susskind, L. (2020) "On the Hardness of Detecting Macroscopic Superpositions," arxiv.org/abs/2009.07450.

[2] science.sciencemag.org/content/early/2020/12/02/science.abe8770.

[3] science.sciencemag.org/content/361/6400/313.

[4] www.nature.com/articles/s41586-019-1666-5.

[5] www.belfercenter.org/publication/public-purpose-consortium-enabling-emerging-technology-public-mission.

## Appendix A

[1] xlinux.nist.gov/dads/HTML/polynomialtm.html.

# Index

**D**



**E**

## I



## J



## K



## L



## M

## N

## R

# About the Editor

**Greg Viggiano, PhD**

Over the course of his career, Greg has had a fascination with observing the benefits and social impacts of using new technologies. Convergence between multiple technologies are especially interesting to him because of the increased potential for unintended effects. These unintended surprises sometimes lead to novel trends and super-trajectories. Much like a digital Swiss Army knife, the smartphone is a perfect example of multi-dimensional convergence across many different uses and applications.

AI and quantum computing offers another interesting research opportunity to examine two technologies that are on a convergent path. Currently, Greg is an adjunct professor at George Mason University, Department of Physics and teaches graduate and undergraduate classes on new technologies and social impacts. His research interests focus on new technology applications and macro social effects. Using a form of complexity theory as a reference framework, his work examines how new technologies are adopted and diffused within global communities—not unlike epidemiological studies of pandemic viral infections. As new technologies become more ubiquitous and evolve [mutate] over time, it is useful to consider how to avoid potential dependencies and vulnerabilities. Using a

10-year time horizon, AI and quantum computing are two such technologies that are coming into alignment.

Science fiction narratives can provide instructive guidance about new technology trends and effects. In addition to his academic career, Greg is also the pro bono Executive Director for the Museum of Science Fiction in Washington, DC. In this context, he studies some of the more prophetic ideas concerning applied uses of fictional technologies. These studies allow greater awareness of what may be waiting over the horizon and glimpses of how civilization might cope with such new arrivals.

To help explore this new horizon and its potential impacts, the Museum of Science Fiction is opening a VR museum in 2023. This virtual reality experience presents several galleries and exhibitions and will begin with "Creating Consciousness: Computers and Robots in Science Fiction." This exhibition examines a machine's capacity to understand humanity and includes a VR tour inside of a quantum computer.

When not teaching and doing social science research on AI and quantum computing, his favorite analog hobbies include European travel photography with his Nikon F and restoring vintage turntables for his ever expanding vinyl record collection. Greg received his PhD from Florida State University in Mass Communication and lives in Alexandria, Virginia with his wife, Mandy.